A

SHORTER   GEOMETRY

CAMBRIDGE UNIVERSITY PRESS
London: FETTER LANE, E.C.
C. F. CLAY, Manager

Edinburgh: 100, PRINCES STREET
Berlin: A. ASHER AND CO.
Leipzig: F. A. BROCKHAUS
New York: G. P. PUTNAM'S SONS
Bombay and Calcutta: MACMILLAN AND Co., Ltd.
Sydney and Melbourne: ANGUS & ROBERTSON.

A
SHORTER GEOMETRY

BY

C. GODFREY, M.V.O., M.A.
HEAD MASTER OF THE ROYAL NAVAL COLLEGE, OSBORNE;
FORMERLY SENIOR MATHEMATICAL MASTER AT WINCHESTER COLLEGE

AND

A. W. SIDDONS, M.A.
LATE FELLOW OF JESUS COLLEGE, CAMBRIDGE; ASSISTANT MASTER
AT HARROW SCHOOL

CAMBRIDGE
AT THE UNIVERSITY PRESS
1912

𝕮𝖆𝖒𝖇𝖗𝖎𝖉𝖌𝖊:

PRINTED BY JOHN CLAY, M.A.

AT THE UNIVERSITY PRESS

PREFACE

THE plan of this book is as follows:

First Stage. Introductory work concerned with the fundamental concepts, and not primarily designed to give facility in using instruments.

Second Stage. Discovery of the fundamental facts of geometry, by experiment and intuition; including the facts relating to angles at a point, parallels, angles of a triangle and polygon, congruent triangles.

When discovered and enunciated, each fact or group of facts is followed by numerical examples and easy riders intended to illustrate and drive home the facts discovered.

In the course of this stage the pupil, besides becoming familiar with the fundamental facts, is taught the accurate use of instruments and the elementary ideas of logical argument as used in strict theoretical geometry.

Third Stage. Deductive development of the propositions subsequent to those dealt with by experiment and intuition in the second stage.

During this stage, experimental and intuitional methods of approaching propositions are by no means excluded. But exercises in drawing are not introduced merely for the sake of drawing; wherever such exercises occur, it is believed that they will help the student to master a geometrical fact, a line of reasoning, or a method of construction.

The division of the geometry course into these three stages is in accordance with the recommendations of the Board of Education (the Teaching of Geometry and Graphic Algebra, Circular 711, March 1909).

The fundamental facts discovered during the Second Stage are not, at this stage, treated as theorems to be proved deductively; the deductive proofs of these fundamental theorems are given in an appendix, and may be postponed till Book IV has been read or till they are needed for examination purposes.

In support of such postponement, the views expressed in the Board of Education Circular may be quoted. There it is argued (i) that the deductive proofs of these propositions present more difficulties, and are less persuasive than the proofs of propositions that follow, e.g. in Book III, (ii) that difficulties of sequence are for the most part confined to these propositions, (iii) that, in the experience of the Board, subsequent work is better in those schools where these propositions are dealt with more rapidly. "These fundamental propositions are those on which all the subsequent deduction depends, and the essential thing in regard to them is not to analyse them, and reduce them to the minimum number of axioms, or, rather, postulates (which is Euclid's method), but to present them in such a way that their truth is as obvious and real to the pupil as the difference between white and black, or between his right hand and his left. Any process which interferes with this directness of vision and apprehension is vicious, whatever claim it may have to logical value, and avenges itself in gross mistakes in subsequent work, due to haziness or lack of grasp of the fundamental facts which have been so laboriously 'proved.'

"With beginners, then, Euclidean proofs of these propositions are out of place, and attention must be concentrated not on formal proofs but on vivid presentation, and accurate, firm apprehension of the propositions themselves."

To justify the title of the present book, we may point out that it is 87 pages shorter than our Elementary Geometry. On the other hand, nothing essential has been omitted. The saving of time to the pupil will be proportionally even greater. Drawing exercises were very numerous in the Elementary Geometry, and a pupil who was made to work anything like the whole of them must have wasted a good deal of time. No one was expected to work them all; but in the present volume we have anxiously guarded against any such waste of time by a severe cutting down of drawing exercises in the Third Stage. We have indicated that in general a neat and good-sized freehand figure should suffice at this stage,—straight lines to look straight, parallel lines parallel and right angles right. Occasionally, of course, there is good reason for an accurate drawing, and this we have required as occasion arose.

The number of theoretical exercises has not been reduced; in fact, it has been increased, notably by a set at the end of Book II.

A possible disadvantage in the new method of teaching geometry is that the pupil may waste a good deal of time in carrying out experiments without a notion of what he is aiming at. On the other hand, if he knows what he is aiming at, experiment may be a sham. The fact is that many of these heuristic exercises are essentially for class work, the class cooperating with the master in developing a new idea, the master giving enough guidance to focus the whole discussion. Exercises that we recommend to be discussed in class are distinguished thus: ¶Ex. 799.

The propositions and their numbers are the same as in the Elementary Geometry, except that the following unnecessary propositions have been omitted: I. 19, 20; IV. 9, 10, 11.

Many teachers (ourselves included) like to take Book III before Book II. The difficulty in arranging a text-book on these lines arises from Pythagoras' theorem, which is needed for numerical exercises in Book III. But this theorem is not referred to in the theorems of Book III: and as the fact involved must be familiar to pupils from their experimental work, there is no good reason why teachers who use this book should not interchange the order of Books II and III.

Another matter in which a good teacher may improve on the text-book is the introduction of three-dimensional work. Occasions arise continually in which the discussion may be lifted out of the plane: e.g. in connection with parallels and perpendiculars, loci, generation of surfaces of revolution, coordinates, etc. Such opportunities will be more likely to suggest themselves if the class room is well furnished with geometrical models.

We are indebted to the courtesy of H.M. Stationery Office, the Admiralty, the Oxford and Cambridge University Presses, the Oxford Local Examinations Delegacy, the Cambridge Local Examinations Syndicate, the Oxford and Cambridge Schools Examination Board and the University of London for permission to reprint certain papers and to use exercises.

C. G.
A. W. S.

December, 1911.

CONTENTS

FIRST STAGE

SECOND STAGE

THIRD STAGE

Book I

Book II. Area

Book III. The Circle

BOOK IV. SIMILARITY

Appendix

The following instruments will be required:—

A hard pencil (HH).

A ruler about 6″ long (or more) graduated in inches and tenths of an inch and also in cm. and mm.

A set square (60°); its longest side should be at least 6″ long.

A semi-circular protractor.

A pair of compasses (with a hard pencil point).

The pencil should have a chisel-point.

The compass pencil may have a chisel-point, or may be sharpened in the ordinary way.

In testing the equality of two lengths or in transferring lengths, compasses should always be used.

Exercises distinguished by a paragraph sign thus: ¶Ex. 15, are intended for discussion in class.

Exercises of a theoretical character (riders) are marked thus: †Ex. 120.

FIRST STAGE.

SOLID.

Take a cylinder, e.g. an uncut lead-pencil. It is called a **geometrical solid**.

Any object, whatever its shape, and whatever it is made of, is called a **solid** in geometry.

Instances of solids in the geometrical sense—the flame of a candle, a brick, a sheet of cardboard or paper, a cloud, a smoke-ring, a drop of water, the air inside a football, etc.

SURFACE.

The pencil is bounded by three **surfaces**. There is the **curved surface**; and there are the flat surfaces at the ends. These ends are called **plane surfaces**, or **planes**.

A solid is bounded by surfaces.

Test for planeness. To test whether a surface is plane, take a straight edge (the edge of a ruler) and apply it to the surface. If the surface is plane, the straight edge should fit in every position.

¶ **Ex. 1.** Test in this way the *curved* surface of a cylinder. You will find that the straight edge fits in some positions, but not in all.

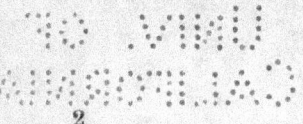

¶Ex. 2. Test in the same way the curved surface (i) of a cone, (ii) of a sphere.

¶Ex. 3. Give instances of solids bounded entirely by plane surfaces.

¶Ex. 4. Give instances of solids bounded by a curved surface.

¶Ex. 5. Give instances of solids bounded partly by one or more plane surfaces, and partly by one or more curved surfaces.

You must not suppose that a leaf of this book is a surface. It is solid; it has thickness. It is bounded by two pages, which *are* surfaces.

A surface has no thickness.

Think of a soap-bubble. The film is not a surface: it is a thin solid, in the shape of a hollow sphere. It *has* an outside surface, and an inside surface. The air inside is in the shape of a geometrical solid, a sphere.

LINE.

The curved surface of the pencil meets the plane end in an edge; this is a curved **line.**

Where two surfaces meet, a line is formed.

If a line referred to is straight, be careful to call it a **straight line.**

¶Ex. 6. Give other instances of two surfaces of a solid meeting (i) in a curved line, (ii) in a straight line.

Neither a piece of fine cotton, nor a fine pencil streak on paper, is, properly speaking, a line. Both of them have thickness and breadth.

A line has neither breadth nor thickness.

POINT.

Take a cube, or cuboid (e.g. a brick). Find two of its edges that meet. They meet at a point.

A point is the meeting-place, or intersection of two lines.

A fine particle of dust, or a pencil dot on paper, is not, strictly speaking, a point.

A point has neither length, breadth nor thickness.

We have now considered in turn a solid, a surface, a line, and a point. We can also consider them in the reverse order.

A point has position but no magnitude.

If a point moves, its path is a line (it is said to generate a line).

A pencil point when moved over a sheet of paper leaves a streak behind, showing the line it has generated (of course it is not really a line, for it has some thickness).

If a line moves, as a rule it generates a surface.

A piece of chalk when laid flat on the blackboard and moved sideways leaves a whitened surface behind it. Consider what would have happened if it had moved along its length.

If a surface moves, as a rule it generates a solid.

The rising surface of water in a dock generates a (geometrical) solid.

¶Ex. 7. Does a flat piece of paper moved along a flat desk generate a solid?

MODELS OF SOLIDS.

The surfaces that bound a solid are generally called **faces**; the lines in which faces meet are called **edges**; the points in which edges meet are called **vertices**.

¶Ex. 8.　Fill up a table such as the following by the help of models.

	Number of faces	Number of edges	Number of vertices
Cube			
Cuboid (or rectangular block)			
3-sided pyramid			
4-sided pyramid			
6-sided pyramid			
3-sided prism			
4-sided prism			
6-sided prism			
Cone			
Cylinder			
Sphere			
Hemisphere			

¶Ex. 9.　Place (or hold) a cube so that you can only see one of its faces; make a sketch of it.

¶Ex. 10. Place (or hold) a cube so that you can see two of its faces; make a sketch of it.

¶Ex. 11. Place a cube so that you can see three of its faces; make a sketch of it.

¶Ex. 12. Can you place a cube so that you can see more than three faces at the same time?

¶Ex. 13. Repeat Exx. 9–12 for a 3-sided prism.

¶Ex. 14. Make sketches of a 6-sided prism in various positions. What is the greatest number of faces you can see at one time?

¶Ex. 15. Make sketches of a 3-sided pyramid in various positions. What is the greatest number of faces you can see at one time?

¶Ex. 16. Make sketches of a 4-sided pyramid in various positions. What is the greatest number of faces you can see at one time?

¶Ex. 17. Make sketches of a cylinder and a cone.

¶Ex. 18. How many measurements are needed to describe the dimensions of a solid, such as a brick?

¶Ex. 19. If you wanted to pack in a packing-case a solid such as a hatbox, or a bookcase, or a tub, how many measurements would you have to make in order to find the necessary dimensions for the case?

¶Ex. 20. How many measurements are needed to describe the dimensions of a surface such as a page of a book?

¶Ex. 21. How many measurements are needed to describe the dimensions of a line such as the edge of a page?

Ex. 22. Make a paper or cardboard box in the shape of a cuboid. You must first draw a pattern such as fig. 1.

Ex. 23. Make a paper or cardboard model of (i) a triangular prism (see fig. 2), (ii) a square pyramid.

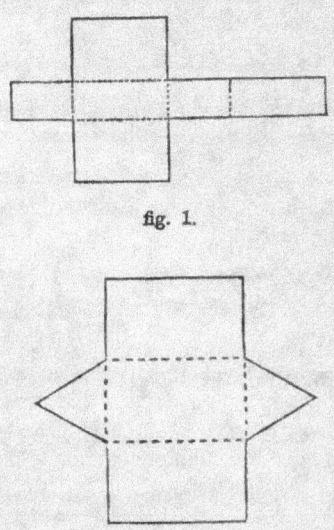

fig. 1.

fig. 2.

MEASUREMENT OF STRAIGHT LINES.

In stating the length of a line, remember to give the unit; the following abbreviations may be used:—*in.* for *inch*; *cm.* for *centimetre*; *mm.* for *millimetre*.

In Exx. 1–49, all lengths measured in inches are to be given to the nearest tenth of an inch, all lengths measured in centimetres to the nearest millimetre.

Always give your answers in decimals.

Ex. 24. Measure the lengths AB, CD, EF, GH in fig. 3

    (i)   in inches,

    (ii) in centimetres.

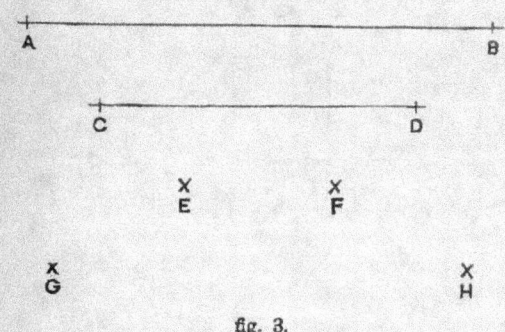

fig. 3.

Ex. 25. Measure in inches and centimetres the lengths of the edges of your wooden blocks.

Ex. 26. Measure in inches the lengths AB, BC, CD in fig. 4; arrange your results in tabular form and add them together.

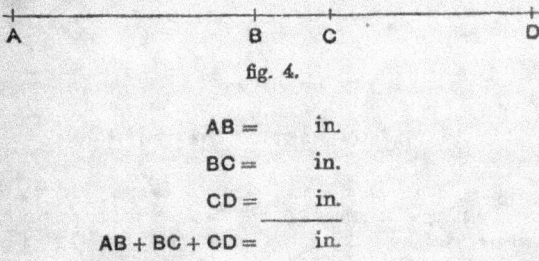

fig. 4.

$$AB = \text{in.}$$
$$BC = \text{in.}$$
$$CD = \text{in.}$$
$$AB + BC + CD = \text{in.}$$

Check by measuring AD.

Ex. 27. Repeat Ex. 26, using centimetres instead of inches.

Ex. 28. Repeat Ex. 26, for fig. 5, (i) using centimetres, (ii) using inches.

X   X   X   X
A   B   C   D

fig. 5.

Ex. 29. Measure in centimetres the lengths AB, BC in fig. 6, and find their difference; arrange your results in tabular form.

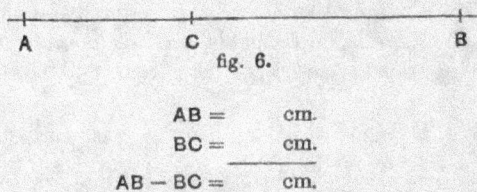

fig. 6.

$$AB = \qquad cm.$$
$$BC = \qquad cm.$$
$$\overline{AB - BC = \qquad cm.}$$

Check by measuring AC.

Ex. 30. Repeat Ex. 29, using inches instead of centimetres.

Ex. 31. Repeat Ex. 29, for fig. 7, (i) using inches, (ii) using centimetres.

X   X   X
A   C   B

fig. 7.

Ex. 32. Measure in inches, and also in centimetres, the length of the paper you are using.

Your ruler is probably too short to measure directly; divide the length into two (or more) parts by making a pencil mark on the edge, and add these lengths together.

Ex. 33. Measure the breadth of your paper in inches and also in centimetres.

Ex. 34. Draw a straight line about 6 in. long and cut off a part AB = 2 in., a part BC = 1·5 in., and a part CD = 1·8 in.; find

the length of AD by adding these lengths; check by measuring
AD. [Make a table as in Ex. 26.]

Ex. 35.   Repeat Ex. 34, with

(i)   AB = 2·7 cm.,   BC = 9·6 cm.,   CD = 1·3 cm.
(ii) AB = 5·2 cm.,   BC = 3·9 cm.,   CD = 2·8 cm.
(iii) AB = ·7 in.,   BC = 2·6 in.,   CD = 2·4 in.
(iv) AB = ·8 cm.,   BC = ·5 cm.,   CD = 2·4 cm.
(v)   AB = 1·8 in.,   BC = 2·9 in.,   CD = ·6 in.

Ex. 36.   A man walks 3·2 miles due north and then 1·5 miles
due south, how far is he from his starting point? Draw a plan
(1 mile being represented by 1 inch) and find the distance by
measurement.

Ex. 37.   A man walks 5·4 miles due west and then 8·2 miles
due east, how far is he from his starting point? (Represent
1 mile by 1 centimetre.)

Ex. 38.   A man walks 7·3 miles due south, then 12·7 miles
due north, then 1·1 miles due south, how far is he from his
starting point? (Represent 1 mile by 1 centimetre.)

Ex. 39.   Draw a straight line, guess its middle point and
mark it by a short cross-line; test your guess by measuring the
two parts.

Ex. 40.   Repeat Ex. 39, three or four times with lines of
various lengths. Show by a table how far you are wrong.

Ex. 41.   Draw a straight line of 10·6 cm.; bisect it by calcu-
lating the length of half the line and measuring off that length
from one end of the line, then measure the remaining part.

When told to draw a line of some given length, you should draw a line
a little too long and cut off a part equal to
the given length as in fig. 8. You should
also write the length of the line against it,
being careful to state the unit.

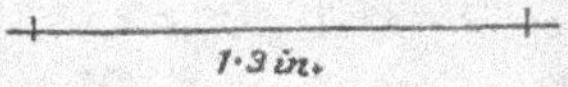

$1·3$ in.

fig. 8.

Ex. 42.   Draw a straight line 3·2 in. long, bisect it as in
Ex. 41.

Ex. 43.   Draw a straight line 2·7 in. long, bisect it as in
Ex. 41.

Ex. 44. Draw straight lines of the following lengths, bisect each of them: (i) 7·6 cm., (ii) 10·5 cm., (iii) 4·1 in., (iv) 9 in., (v) 5·8 cm., (vi) 11·3 cm.

A good practical method of bisecting a straight line (AB) is as follows (see fig. 9):—measure off with compasses equal lengths (AC, BD) from each end of the line (these lengths should be very nearly half the length of the line) and bisect the remaining portion (CD) by eye.

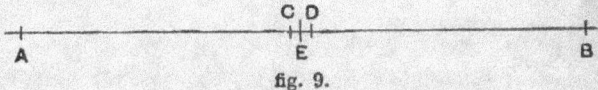

fig. 9.

Ex. 45. Draw three or four straight lines and bisect them with your compasses (as explained above); verify by measuring each part of the line (remember to write its length against each part).

Ex. 46. Open your compasses 1 cm., apply them to the inch scale and so find the number of inches in 1 centimetre.

Ex. 47. Find the number of inches in 10 cm. as in Ex. 46; hence express 1 cm. in inches. Arrange your results in tabular form.

Ex. 48. Find the number of centimetres in 5 in. as in Ex. 46; hence find the number of centimetres in 1 inch.

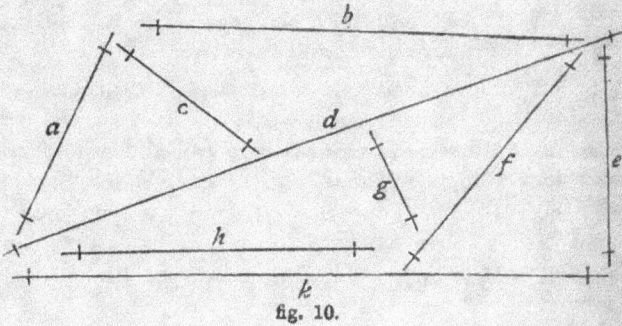

fig. 10.

Ex. 49. Guess the lengths of the lines in fig. 10 (i) in inches, (ii) in centimetres; verify by measurement. Make a table thus :—

Line	Guessed	Measured
a		
b		

DIRECTION.

¶Ex. 50. Point vertically upwards.

¶Ex. 51. Point vertically downwards.

¶Ex. 52. Show me some vertical lines in the room.

¶Ex. 53. How would you test whether the edge of an open door is truly vertical?

¶Ex. 54. Show me three or four horizontal lines.

¶Ex. 55. How would you test whether a line is truly horizontal?

¶Ex. 56. Place a cube with *one* edge vertical. How many edges do you now find to be vertical? How many horizontal?

¶Ex. 57. Place a cube with one edge horizontal, and no edge vertical. How many edges are now horizontal?

¶Ex. 58. Place a cube with one edge vertical, and no edge horizontal.

¶Ex. 59. How many vertical lines can you have through a point? how many horizontal?

¶Ex. 60. How many horizontal lines can you draw on a sloping desk through a point?

¶Ex. 61. How many vertical lines can you draw on a sloping desk?

¶Ex. 62. How many horizontal lines can you draw on the wall through a point? how many vertical?

¶Ex. 63. How many horizontal lines can you draw on the floor through a point? how many vertical?

¶Ex. 64. Show me two or three horizontal planes? Are they parallel? Show me a line at right angles to a horizontal plane. What is its direction called?

¶Ex. 65. Show me two or three vertical planes. Are they parallel? Show me a line at right angles to a vertical plane. What is its direction called?

¶Ex. 66. Point in a Northward direction.

¶Ex. 67. Point in an Eastward direction.

¶Ex. 68. Two boys stand out and face in the same direction.

¶Ex. 69. All the class point in the same direction, say West. Notice that all the arms are parallel.

¶Ex. 70. All look at the same point on the blackboard. Are all looking in the same direction?

¶Ex. 71. Give some other instances of lines, not horizontal, in the same direction.

¶Ex. 72. Are all vertical lines in the room in the same direction?

¶Ex. 73. Are all horizontal lines in the room in the same direction?

¶Ex. 74. Two men are looking at the same star. Are they looking in the same direction?

¶Ex. 75. Two men are looking at the same weathercock. Are they looking in the same direction?

ANGLE.

¶Ex. 76. What angle* do you turn through on receiving the following orders: (i) right turn, (ii) left turn, (iii) about turn, (iv) half right turn?

¶Ex. 77. What angle* is there between the following directions: (i) N. and N.E., (ii) N. and E., (iii) N. and S.E., (iv) N. and S.? [*In each case, draw the directions freehand.*]

¶Ex. 78. What is the angle* between N.E. and S.W.?

¶Ex. 79. How long does the earth take to turn through a right angle?

¶Ex. 80. Through how many right angles does the hour hand of a clock turn in 3 hours, 1 hour, 6 hours, 9 hours?

¶Ex. 81. A yacht sails round the course marked in fig. 11. Copy the figure freehand, mark ∠s turned through at each corner, and guess their sizes*.

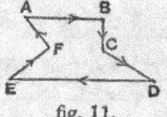

fig. 11.

Ex. 82. Repeat Ex. 81 for fig. 12.

Ex. 83. Repeat Ex. 81 for fig. 13.

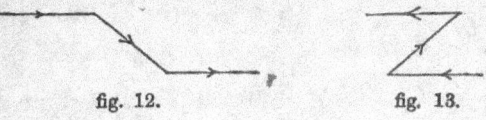

fig. 12.　　　　　　　　fig. 13.

¶Ex. 84. Through what angle (to the nearest half right angle) has the batsman turned the direction of the ball when he sends it to (i) point, (ii) long slip, (iii) cover point, (iv) mid on?

¶Ex. 85. Through what angle* is the direction of the billiard ball A turned, when it cannons off B on to C (fig. 14)?

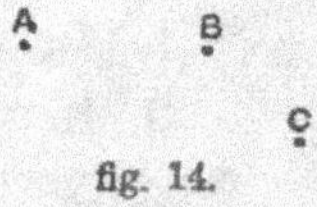

fig. 14.

* Answer to be in right angles and fractions of a right angle

¶Ex. 86. The wind shifts from W. to N.E. Through what angle* has it turned?

Ex. 87. A man walks one mile N., then turns through $\frac{1}{2}$ rt ∠ to the right and walks another mile. Draw his path, guessing the angle. (Scale 1 inch to 1 mile.)

Ex. 88. Draw the following journey to scale. One mile N., then turn through $\frac{1}{2}$ rt ∠ left and go 2 miles; then turn $1\frac{1}{2}$ rt ∠ s left and go 1 mile; then turn $\frac{1}{2}$ rt ∠ left and go 2 miles. (Scale 1 inch to 1 mile.)

¶Ex. 89. A boy sits on a horse on a merry-go-round. When the machine has made one round, what angle* has the boy turned through?

If you hold one arm of your compasses firm and turn the other about the hinge, the two arms form an angle.

In the same way if two straight lines OA, OB are drawn from a point O, they form an

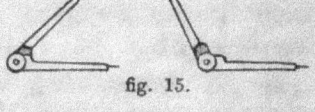

fig. 15.

angle at O. O is called the **vertex** of the angle, and OA, OB its **arms**.

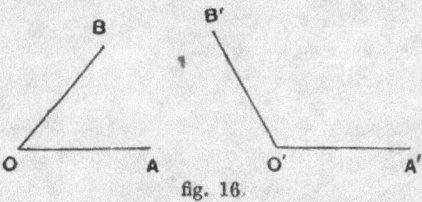

fig. 16.

An angle may be denoted by three letters; thus we speak of the angle AOB, the middle letter denoting the vertex of the angle and the outside letters denoting points on its arms.

* Answer to be in right angles and fractions of a right angle.

If there is only one angle at a point O, we call it the angle O.

Sometimes an angle is denoted by a small letter placed in it; thus in fig. 17 we have two angles *a* and *b*.

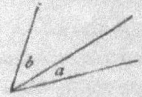

fig. 17.

∠ is the abbreviation for *angle*.

Two angles AOB, CXD (see figs. 16 and 18), are said to be **equal** when they can be made to fit on one another exactly (i.e. when they are such that, if CXD be cut out and placed so that X is on O and XC along OA, then XD is along OB). It is important to notice that it is not necessary for the *arms* of the one angle to be equal to those of the other, in fact *the size of an angle does not depend on the lengths of its arms.*

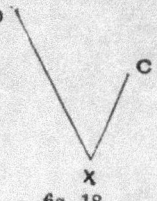

fig. 18.

¶Ex. 90. Draw an angle on your paper and open your compasses to the same angle.

¶Ex. 91. Which is the greater angle in fig. 19? Test by making on tracing paper an angle equal to one of the angles and fitting the trace on the other.

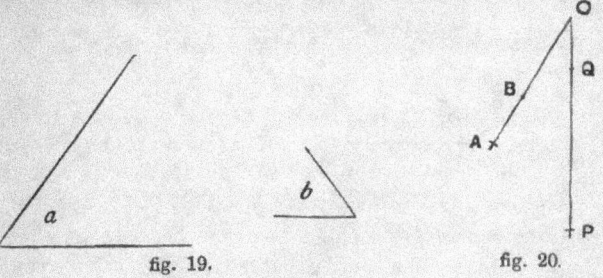

fig. 19.   fig. 20.

¶Ex. 92. Name the angle at O in fig. 20 in as many different ways as you can.

¶Ex. 93. Take a piece of paper and fold it; you will get
something like fig. 21; fold it
again so that the edge OB fits
on the edge OA; now open the
paper; you have four angles
made by the creases, as in
fig. 22; they are all equal, for
when folded they fitted on
one another. Such angles are
right angles. An angle less
than a right angle is called
an acute angle. An angle
greater than a right angle is
called **an obtuse** angle.

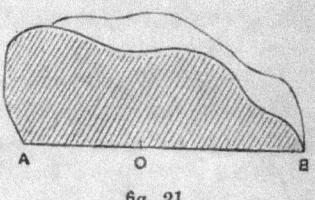

fig. 21.

¶Ex. 94. Make a right
angle BOC as in Ex. 93; cut
it out and fold so that OB
falls on OC. Does the crease
(OE) bisect ∠ BOC? (i.e. are
∠s BOE, EOC equal?) What
fraction of a right angle is
each of the ∠s BOE, EOC?

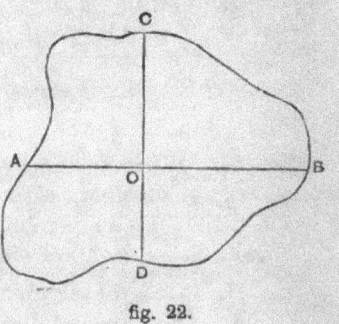

fig. 22.

¶Ex. 95. If the ∠ BOE of Ex. 94 were bisected by folding,
what fraction of a right angle would be obtained?

If a right angle is divided into 90 equal angles, each of these
angles is called a **degree.**

25° is the abbreviation for "25 degrees."

Fig. 23 represents a **protractor**; if each graduation on the edge were joined to C, we should get a set of angles at C each of which would be an angle of one degree

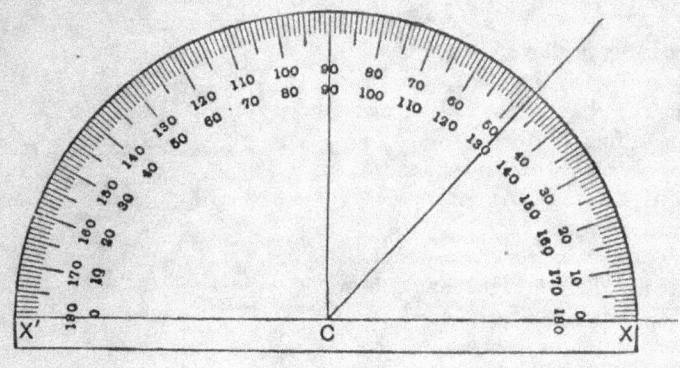

fig. 23.

¶Ex. 96. What fractions of a right angle are the angles between the hands of a clock at the following times:—(i) 3.0, (ii) 1.0, (iii) 10.0, (iv) 5.0, (v) 8.0? State in each case whether the angle is acute, right, or obtuse.

Ex. 97. Find the number of degrees in each of the angles in Ex. 96. [Use the results of that Ex.]

Ex. 98. Fig. 24 shows the points of the compass; what are the angles* between (i) N and E, (ii) W and S W, (iii) W and W N W, (iv) E and E by S, (v) N E and N N W, (vi) S W and S E?

To measure an angle, place the protractor so that its centre C is at the vertex of the angle and its base, CX, along one arm of the angle; then note under which graduation the other arm passes; thus in fig. 23, the angle = 48°.

In using a protractor such as that in fig. 23, care must be taken to choose the right set of numbers; e.g. if the one arm of the angle to be measured

* Answers to be in degrees.

lies along CX, the set of numbers to be used is obviously the one in which
the numbers increase as the line turns round C from CX towards CX'.

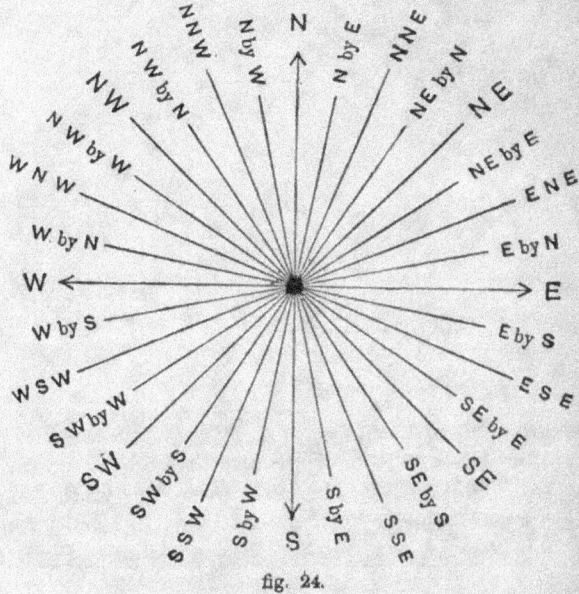

fig. 24.

You should also check your measurement by noticing whether the angle
is acute or obtuse.

When you measure an angle in a figure that you have
drawn (or make an angle to a given measure), always indicate
in your figure the number of degrees, as in fig. 25.

¶Ex. 99. Cut out of paper a right angle, bisect it
by folding, and measure the two angles thus formed.   fig. 25.

¶Ex. 100.   Measure the angles of your set square (i) directly,
(ii) by making a copy on paper and measuring the copy.

It is difficult to draw a straight line right to the corner of a set square;
it is better to draw the lines to within half a centimetre of the corner and
afterwards produce them (i.e. prolong them) with the ruler till they meet.

¶Ex. 101. Measure the angles of your models—this may be done either directly, or more accurately by copying the angles and measuring the copy.

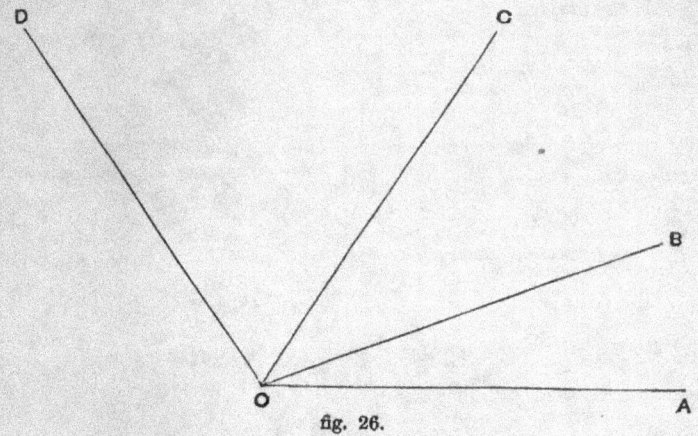

fig. 26.

Ex. 102. Measure ∠s AOB, BOC in fig. 26; add; and check your result by measuring ∠ AOC. (Arrange in tabular form.)

Ex. 103. Measure ∠s AOC, COD, AOD in fig. 26. Check your results.

Ex. 104. Measure ∠s AOB, BOD, AOD in fig. 26. Check your results.

Ex. 105. Repeat the last three exercises for fig. 28.

Ex. 106. Draw a circle (radius about 2·5 in.), cut off equal parts from its circumference (this can be done by stepping off with compasses or dividers). Join OA, OB, OF. Measure ∠s AOB, AOF. Is ∠ AOF = 5 times ∠ AOB?

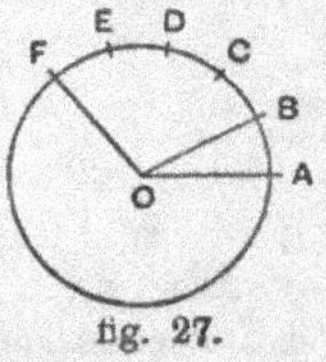

fig. 27.

To make an angle to a given measure. Suppose that you have a line AB and that at the point A you wish to make an angle of 73°. Place the protractor

2—2

so that its centre is at A and its base along AB, mark the 73°
graduation with the point of your compasses (only a small prick
should be made), and join this point to A.   (Remember to write
73° in the angle.)

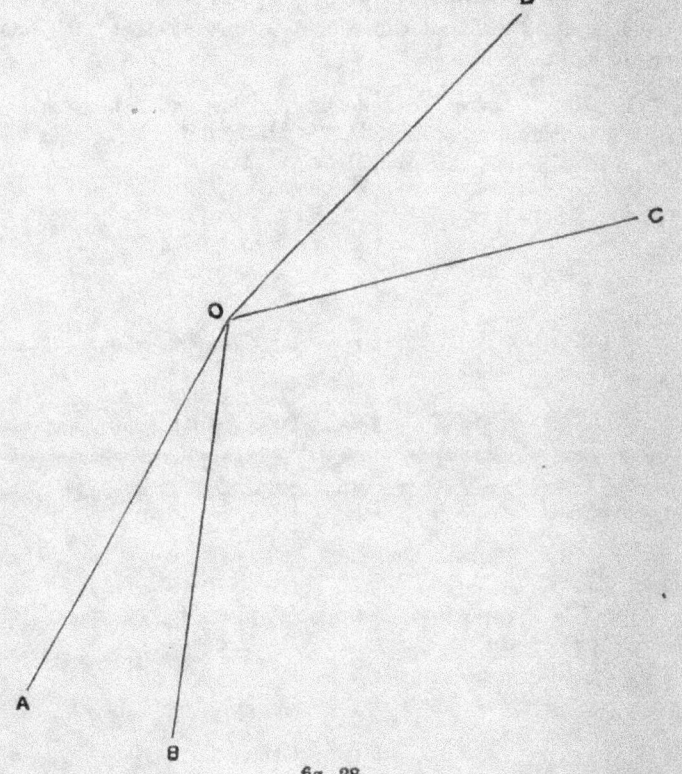

fig. 28.

Ex. 107. Make a copy of the smallest angle of your set
square and bisect it as follows :—measure the angle with your
protractor, calculate the number of degrees in half the angle,

mark off this number (as explained above) and join to the vertex. Verify by measuring each half. (This will be referred to as the method of bisecting an angle *by means of the protractor.*)

Ex. 108. Make angles of 20°, 35°, 64°, 130°, 157°, 176° (let them point in different directions). State whether each one is acute, right, or obtuse.

Ex. 109. Make the following angles and bisect each by means of the protractor, 24°, 78°, 152°, 65°, 111°. (Let them point in different directions.)

In the Second Stage exercises leading up to new facts are printed in ordinary type; exercises designed to give practice in the application of these facts are in small type. Exercises of a theoretical character (riders) are marked thus (†).

SECOND STAGE.

ANGLES AT A POINT.

¶Ex. 110.   Draw an acute angle AOB; produce
AO to C; what kind of angle is BOC? (*Freehand*).

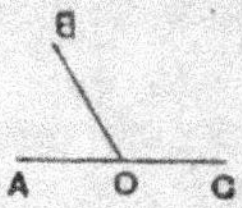

¶Ex. 111.   Draw an obtuse angle BOC; produce
CO to A; what kind of angle is AOB? (*Freehand*).

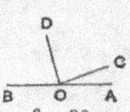

fig. 29.

¶Ex. 112.   (i) Make ∠AOB = 65°; produce AO to C; measure
∠BOC; what is the sum of ∠s AOB, BOC?

(ii)   Repeat (i) with ∠AOB = 77°.

(iii)   Repeat (i) with ∠AOB = 123°.

Compare the results of (i), (ii), (iii); how many right angles
are there in each sum?

¶Ex. 113.   From a point O in a straight line
AB, draw two lines OC, OD (see fig. 30); measure
the three angles; what is their sum?

fig. 30.

¶Ex. 114.   Repeat Ex. 113, with AB drawn in
a different direction.

¶Ex. 115.   In fig. 29, suppose a line to rotate about O from
the position OA, through the position OB to the position OC.
(i) Through how many right angles has the line rotated?
(ii) Through how many degrees?

¶Ex. 116.   What do you infer from the above exercises?

23

FACT A₁. [Theorem I. 1.]

If a straight line stands on another straight line, the sum of the two angles so formed is equal to two right angles.

Ex. **117.** If, in fig. 29, ∠ AOB = 57°, what is ∠ BOC? Check by drawing and measuring.

¶Ex. **118.**   (i)   If, in fig. 29, ∠ BOC = 137°, what is ∠ AOB?

(ii)   ,,   ,,   ∠ BOC = 93°   ,, ,, ∠ AOB?

(iii)   ,,   ,,   ∠ AOB =   5°   ,, .. ∠ BOC?

¶†Ex. **119.** If two straight lines AOB, COD intersect at O and ∠ AOC is a right angle, prove that the other angles at O are right angles.

¶†Ex. **120.** If △ ABC has ∠ ABC = ∠ ACB, prove that the exterior angles formed by producing the base both ways must be equal to one another (i.e. prove that ∠ ABD = ∠ ACE)*. See fig. 31.

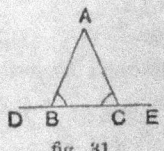

fig. 31.

†Ex. **121.** In △ ABC, ∠ ABC = ∠ ACB and AB and AC are produced to X and Y, prove that ∠ CBX = ∠ BCY. (See fig. 32.)

DEF. When the sum of two angles is equal to two right angles, each is called the **supplement** of the other, or is said to be **supplementary** to the other.

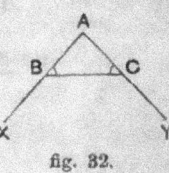

fig. 32.

¶Ex. **122.** What are the supplements of the following angles, (i) 60°, (ii) 54°, (iii) 0°, (iv) 90°, (v) 180°, (vi) x°?

¶Ex. **123.** Name the supplements of ∠ ABC and ∠ BCY in fig. 32.

Name the supplements of ∠ ABC and ∠ ECA in fig. 31.

Ex. **124.** State Fact A₁, introducing the term "supplementary."

Ex. **125.** In fig. 29, show how to obtain another supplement of ∠ AOB.

---

* When angles or lines are given or made equal, it is well to indicate the fact in your figure by putting the same mark in each.

¶Ex. 126. Draw a straight line OB; on opposite sides of OB make the two angles AOB = 42°, BOC = 129°. What is their sum? Is AOC a straight line?

¶Ex. 127. Repeat Ex. 126, with

 (i) ∠ AOB = 42°, ∠ BOC = 138°.

 (ii) ∠ AOB = 90°, ∠ BOC = 90°.

 (iii) ∠ AOB = 73°, ∠ BOC = 113°.

 (iv) ∠ AOB = 113°, ∠ BOC = 76°.

¶Ex. 128. What connection must there be between the two angles in the last Ex. in order that AOC may be straight?

DEF. When three straight lines are drawn from a point, if one of them is regarded as lying between the other two, the angles which this line makes with the other two are called **adjacent** angles (e.g. ∠ˢ *a* and *b* in fig. 17).

¶Ex. 129. State what fact you infer from the above exercises.

FACT B. [Theorem I. 2.]

If the sum of two adjacent angles is equal to two right angles, the exterior arms of the angles are in the same straight line.

Ex. 130. Make an ∠ AOB = 36°; produce AO to C; make ∠ COD = 36° as in fig. 33; calculate ∠ BOC; is BOD a straight line in your figure? Give a reason.

†Ex. 131. From a point A in a straight line AB, straight lines AC and AD are drawn at right angles to AB on opposite sides of it; prove that CAD is a straight line.

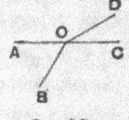

fig. 33.

†Ex. 132. From a point O in a straight line AOC, OB and OD are drawn on opposite sides of AC so that ∠ AOB = ∠ COD (see fig. 33); prove that BOD is a straight line.

DEF. If a straight line or angle is divided into two equal parts it is said to be **bisected.**

†Ex. 133. Two straight lines XOX', YOY' intersect at right angles; OP bisects ∠ XOY, OQ bisects ∠ X'OY'. Is POQ a straight line?

[Find the sum of ∠ˢ POY, YOX', X'OQ.]

¶Ex. 134.   Draw fig. 34 making ∠ BOC = 67° and ∠ B'O'D' = 29°. What is the sum of the four angles?

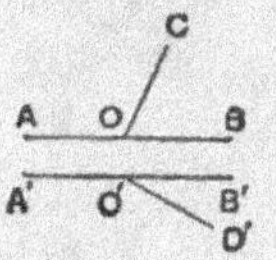

fig. 34.

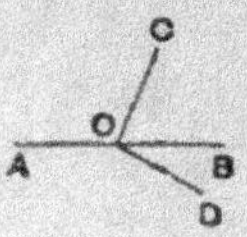

fig. 35.

¶Ex. 135.   Draw fig. 35 making ∠ BOC = 67° and ∠ BOD = 29°. What is the sum of the four angles at O?   Give a reason.

¶Ex. 136.   From a point O in a straight line AB, draw straight lines OC, OD, OE, OF, OG as in fig. 36.   Measure the angles AOC, COD, &c.   What is their sum?

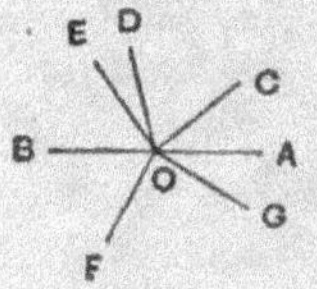

fig. 36.

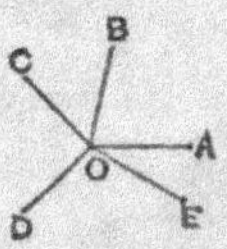

fig. 37.

¶Ex. 137.   From a point O, draw a set of straight lines as in fig. 37; measure the angles so formed.   What is their sum? How many right angles is the sum equal to?

¶Ex. 138.   If in fig. 37 a line rotates about O from the position OA, through the positions OB, OC, OD, OE, back to the position OA, (i) through how many right angles does it rotate? (ii) through how many degrees?

¶Ex. 139.   What fact do you infer from the above exercises?

FACT A₂.   [Theorem I. 1.   Corollary.]

If any number of straight lines meet at a point, the sum of all the angles made by the consecutive lines is equal to four right angles.

¶Ex. 140.　Draw two straight lines as in fig. 38; measure all the angles.

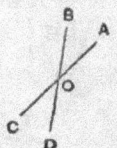

¶Ex. 141.　Make ∠ AOB = 47°; produce AO to C and BO to D; measure all the angles.

¶Ex. 142.　Repeat Ex. 141 with ∠ AOB = 166°.

fig. 38.

¶Ex. 143.　In fig. 38, if ∠ AOB = 73°, what are the remaining angles?

¶Ex. 144.　(i) In fig. 38, if ∠ AOD = 132°, what are the remaining angles?

(ii)　In fig. 38, if ∠ COD = 58°, what are the remaining angles?

(iii)　In fig. 38, if ∠ BOC = 97°, what are the remaining angles?

¶Ex. 145.　Place two pencils along one another; now rotate one of them about its middle point so as to form a letter X; do the two parts of the pencil turn through equal angles?

If the pencil rotates about any other point, will the angles be equal?

DEF.　The opposite angles made by two intersecting straight lines are called **vertically opposite angles** (*vertically* opposite because they have the same vertex).

¶Ex. 146.　Name two pairs of vertically opposite angles in fig. 38.

¶Ex. 147.　State a new fact to be discovered from the above exercises.

FACT C. [Theorem I. 3.]

If two straight lines intersect, the vertically opposite angles are equal.

¶Ex. 148. Draw a triangle and produce every side both ways; number all the angles in the figure, using the same numbers for angles that are equal.

Ex. 149. If two straight lines AOB, COD intersect at O (see fig. 38) what is the sum of ∠ˢ AOB, BOC? What is the sum of ∠ˢ BOC, COD?
Hence prove Fact C.

†Ex. 150. In fig. 39, prove that

   (i) if ∠b = ∠f, then ∠c = ∠f.

   (ii) if ∠c = ∠f, then ∠d = ∠e.

   (iii) if ∠d + ∠f = 2 rt ∠ˢ, then ∠b = ∠f.

   (iv) if ∠g = ∠c, then ∠d = ∠h.

   (v) if ∠h = ∠a, then ∠e = ∠d.

   (vi) if ∠a = ∠e, then ∠b = ∠g.

   (vii) if ∠c = ∠f, then ∠d + ∠f = 2 rt ∠ˢ.

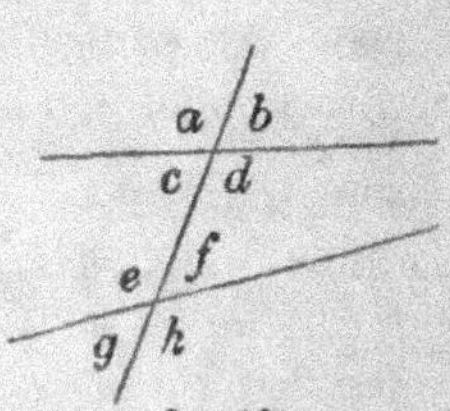

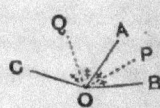

fig. 39.

†Ex. 151. If two angles are equal, their supplements are equal.

†Ex. 152. If two straight lines AOC, BOD intersect at O and OX bisects ∠AOB, then XO produced bisects ∠COD.

†Ex. 153. **The bisectors of a pair of vertically opposite angles are in one and the same straight line.**

Ex. 154. If in fig. 29, ∠BOA = 50°, and OP is drawn bisecting ∠BOA and OQ bisecting ∠AOC; what are ∠ˢ POA, AOQ? What is their sum? (*Freehand*).

†Ex. 155. Three straight lines OA, OB, OC are drawn from a point O (see fig. 40); OP is drawn bisecting ∠BOA, and OQ bisecting ∠AOC. Prove that

    ∠POQ = ½ ∠BOC.

fig. 40.

†Ex. 156. **If a straight line stands on another straight line, prove that the bisectors of the two adjacent angles so formed are at right angles to one another.** (See Exx. 154, 155.)

†Ex. 157. Three straight lines OB, OA, OC are drawn from a point (see fig. 40), OP bisects ∠BOA, OQ bisects ∠AOC; prove that, if ∠POQ is a right angle, BOC is a straight line.

PARALLEL STRAIGHT LINES.

¶Ex. 158. Give instances of parallel straight lines (e.g. the flooring boards of a room, the edges of your paper).

¶Ex. 159. Draw with your ruler two straight lines as nearly parallel as you can judge; draw several lines at right angles to one of the parallels as in fig. 41; measure these lines.

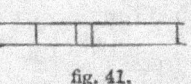

fig. 41.

State what you infer from this experiment.

¶Ex. 160. Draw with your ruler two straight lines as nearly parallel as you can judge; draw a straight line cutting them as in fig. 42; measure the angles marked. These are called **corresponding** angles. Are they equal?

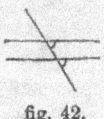

fig. 42.

¶Ex. 161. Name some pairs of corresponding angles in fig. 48.

¶Ex. 162. Place your set square so that one of its edges touches your ruler as at (i) in fig. 43; hold the ruler firm and slide the set square along the ruler to position (ii). Draw the line AB for two positions of the set square; notice that the two straight lines point in the same direction; in fact, the two lines are parallel.

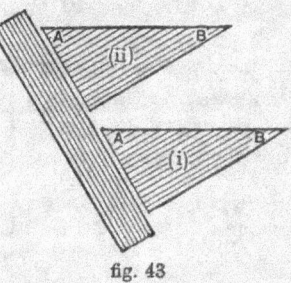

fig. 43

This method of drawing parallels suggests an explanation of the term *corresponding* angles.

¶Ex. 163. Draw a pair of parallel lines as in Ex. 162; draw a line cutting them as in fig. 42; measure the corresponding angles. Repeat this for two or three different positions of the cutting line.

¶Ex. 164. Draw two straight lines which are *not* parallel and proceed as in Ex. 163. Are the angles equal?

¶Ex. 165. Draw fig. 44; first draw PQRS, then make the angles as shown. Test as in Ex. 159 whether QX, RY are parallel.

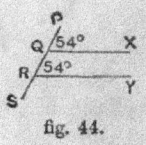

fig. 44.

A man walks along a zig-zag road ABCD, AB being parallel to CD, so that he finishes up by walking in his original direction.

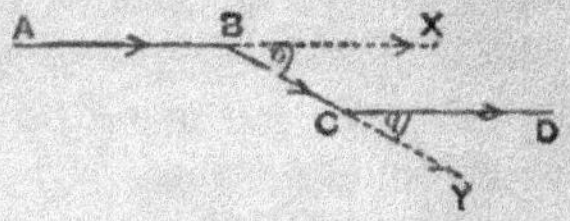

fig. 45.

At B he turns to the *right* through ∠p (N.B. *not* through ∠ABC); and at C he turns to the *left* through ∠q.

Since his last direction is the same as his first, his right and left turns must compensate; ∴ ∠p = ∠q.

Hence we may see that BCY, meeting the parallel lines BX, CD, makes the corresponding angles p and q equal.

From the above and Exx. 160–165 we arrive at a geometrical fact which we will state in two different ways.

FACT D (*corresponding angles*). [Theorem I. 4 (2).]

When a straight line cuts two other straight lines, if a pair of corresponding angles are equal, then the two straight lines are parallel.

FACT E (*corresponding angles*). [Theorem I. 5 (2).]

When a straight line cuts two parallel straight lines, the corresponding angles are equal.

¶Ex. 166. Draw a pair of parallel lines (using set square and ruler as in Ex. 162); draw a line cutting them as in fig. 46. Measure the angles marked. These are called **alternate** angles. Are they equal?

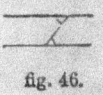

fig. 46.

¶Ex. 167.   Measure another pair of alternate angles in your figure for the last Ex.

¶Ex. 168.   Name some pairs of alternate angles in fig. 48.

¶Ex. 169.   Repeat Ex. 166 for two or three different positions of the cutting line.

¶Ex. 170.   State a fact like Fact D (corresponding angles) for alternate angles.

¶Ex. 171.   State a fact like Fact E (corresponding angles) for alternate angles.

FACT D (*alternate angles*).   [Theorem I. 4 (i).]

When a straight line cuts two other straight lines, if a pair of alternate angles are equal, then the two straight lines are parallel.

FACT E (*alternate angles*).   [Theorem I. 5 (i).]

When a straight line cuts two parallel straight lines, the alternate angles are equal.

¶Ex. 172.   Draw a pair of parallel lines (using set square and ruler); draw a line cutting them as in fig. 47.   Measure the two interior angles on one side of the cutting line.   What is their sum?

fig. 47.

¶Ex. 173.   Measure the two interior angles on the other side of the cutting line.   What is their sum?

¶Ex. 174.   Repeat Ex. 172 for two or three different positions of the cutting line.

¶Ex. 175.   State a fact like Fact D (corresponding angles) for interior angles.

¶Ex. 176.   State a fact like Fact E (corresponding angles) for interior angles.

FACT D (*interior angles*). [Theorem I. 4 (iii).]

When a straight line cuts two other straight lines, if a pair of interior angles on the same side of the cutting line are together equal to two right angles, then the two straight lines are parallel.

FACT E (*interior angles*). [Theorem I. 5 (iii).]

When a straight line cuts two parallel straight lines, the interior angles on the same side of the cutting line are together equal to two right angles.

Ex. 177. Copy fig. 48 freehand. If CD, EF are parallel and ∠AGD=72°, find all the angles in the figure, giving your reasons; make a table of values.

Ex. 178. Repeat Ex. 177 with ∠EHB = 120°, CD, EF being parallel.

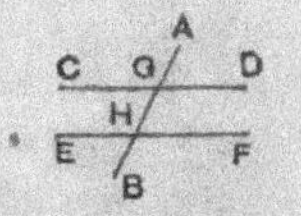

fig. 48.

†**Ex. 179.** In fig. 49 there are two pairs of parallel lines; prove that the following pairs of angles are equal :—(i) *b*, *l*, (ii) *f*, *k*, (iii) *m*, *s*, (iv) *f*, *h*, (v) *r*, *l*, (vi) *s*, *h*, (vii) *s*, *q*, (viii) *s*, *k*, (ix) *s*, *a*, (x) *g*, *l*.

[State your reasons carefully.

e.g. WX, YZ are ∥ and ST cuts them,

∴ ∠*q* = ∠*f* (corresponding angles).]

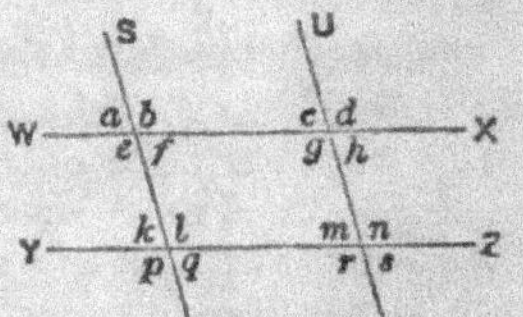

fig. 49.

Ex. 180. What do you know about the sums of (i) ∠ˢ *f*, *g*, (ii) ∠ˢ *f*, *l*, (iii) ∠ˢ *m*, *n*, in fig. 49? Give your reasons.

†**Ex. 181.** In fig. 48 prove that, if ∠AGD = ∠GHF, then ∠CGH = ∠GHF. Prove this (i) using only Facts D, E; (ii) without using Facts D, E.

†**Ex. 182.** In fig. 48 prove that, if ∠CGH = ∠GHF, then ∠AGD = ∠GHF. Prove this (i) using only Facts D, E; (ii) without using Facts D, E.

†**Ex. 183.** Prove that if a straight line cuts two other straight lines and makes a pair of alternate angles equal to one another, then the interior angles on the same side of the cutting line are together equal to two right angles.

Use Facts D, E.

†**Ex. 184.** Repeat Ex. 183 without using Facts D, E.

DEF. A plane figure bounded by four straight lines is called a quadrilateral.

DEF. The straight lines which join opposite corners of a quadrilateral are called its diagonals.

DEF. A quadrilateral with its opposite sides parallel is called a parallelogram.

Ex. 185. Draw a parallelogram ABCD, join AC, and produce BC to E; what pairs of angles in the figure are equal? Give your reasons.

†Ex. 186 ABCD is a quadrilateral, its diagonal AC is drawn; prove that, if ∠BAC=∠ACD and ∠DAC=∠ACB, ABCD is a parallelogram.

†Ex. 187. The opposite angles of a parallelogram are equal. [See Ex. 185.]

†Ex. 188. A triangle ABC has ∠B=∠C, and DE is drawn parallel to BC; prove that ∠ADE=∠AED. See fig. 50.

†Ex. 189. If a straight line is perpendicular to one of two parallel straight lines, it is also perpendicular to the other.

fig. 50.

†Ex. 190. If each of two straight lines is perpendicular to a third straight line, the two straight lines are parallel to one another.

†Ex. 191. What is the sum of the angles of a parallelogram? Hence find the sum of the angles of a triangle.

†Ex. 192. If one angle of a parallelogram is a right angle, prove that all its angles must be right angles.

ANGLES OF A TRIANGLE.

DEF. A plane figure bounded by three straight lines is called a triangle.

¶Ex. 193. Draw any triangle, and measure its angles. What is the sum of its angles?

¶Ex. 194. Repeat Ex. 193 for two or three different triangles.

¶Ex. 195. Cut out a paper triangle; mark its angles; tear off the corners and fit them together with their vertices at one point, as in fig. 51.

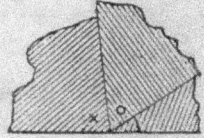

fig. 51.

What relation between the angles of a triangle is suggested by this experiment?

¶Ex. 196. State the fact that you infer from the above exercises?

Fact F_1. [Theorem I. 8.]

The sum of the angles of a triangle is equal to two right angles.

We will now *prove* this fact in Exx. 197, 198.

†Ex. 197. Draw a triangle ABC, produce BC to D, and draw CE parallel to BA. What pairs of angles in the figure are equal? What do you deduce from this about the sum of the angles of a triangle?

†Ex. 198. Draw a triangle ABC; through A draw DAE parallel to BC. Hence find the sum of the angles of the triangle.

¶Ex. 199. Would it be possible to have triangles with angles of
(i)   90°, 60°, 30°,   (ii) 77°, 84°, 20°,   (iii) 59°, 60°, 61°,
(iv) 135°, 22°, 22°,   (v) 73°, 73°, 33°,   (vi) 54°, 54°, 72°?

¶Ex. 200. (i) Give two sets of angles which would do for the angles of a triangle.
(ii) Give two sets which would not do.

¶Ex. 201. If two angles of a triangle are 54°, 76°, what is the third angle?

¶Ex. 202. If two angles of a triangle are 27°, 117°, what is the third angle?

¶Ex. 203. If two angles of a triangle are 23°, 31°, what is the third angle?

¶Ex. 204. If two angles of a triangle are 65°, 132°, what is the third angle?

¶Ex. 205. If the angles of a triangle are all equal, what is the number of degrees in each?

Ex. 206. If one angle of a triangle is 36°, and the other two angles are equal, find the other two angles.

Ex. 207. Repeat Ex. 206 with the given angle (i) 90°, (ii) 132°, (iii) 108°.

Ex. 208. In fig. 52, triangle ABC has ∠A = 90°; AD is drawn perpendicular to BC. If ∠B = 27°, find the angles marked x, y, z.

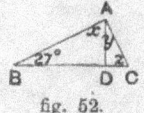

fig. 52.

Ex. 209. Repeat Ex. 208 with (i) ∠B = 54°, (ii) ∠B = 33°, (iii) ∠B = 45°.

Ex. 210. A triangle ABC has ∠A = 75°, ∠B = 36°; if AD is drawn perpendicular to BC, find each angle in the figure.

¶Ex. 211. Draw any quadrilateral and find the sum of its angles.

¶Ex. 212. Cut out a paper quadrilateral and proceed as in Ex. 195.

†Ex. 213. Prove that the sum of the angles of a quadrilateral is equal to four right angles. (Draw a diagonal.)

†Ex. 214. What is the sum of the angles of a pentagon (a five-sided figure)?
[Join one vertex to the two opposite vertices.]

Ex. 215. In a quadrilateral ABCD, ∠A = 77°, ∠B = 88°, ∠C = 99°; find ∠D. (*Freehand*).

Ex. 216. In a quadrilateral ABCD, ∠A = 37°, ∠B = 111°, and ∠C = ∠D; find ∠C and ∠D. (*Freehand*).

†Ex. 217. If one angle of a triangle is a right angle, the other two angles must be acute.

†Ex. 218. If one angle of a triangle is obtuse, the other two angles must be acute.

DEF. A triangle which has one of its angles an obtuse angle is called an **obtuse-angled** triangle.

DEF. A triangle which has one of its angles a right angle is called a **right-angled** triangle.

The side opposite the right angle is called the **hypotenuse**.

DEF. A triangle which has *all* its angles acute angles is called an **acute-angled** triangle.

In Exx. 217, 218, we have seen that **every triangle must have at least two of its angles acute.**

3—2

DEF. A triangle which has two of its sides equal is called an isosceles triangle.

DEF. A triangle which has all its sides equal is called an equilateral triangle.

DEF. A triangle which has no two of its sides equal is called a scalene triangle.

DEF. A triangle which has all its angles equal is said to be equiangular.

Ex. 219. Make a table showing in column A whether the triangles in fig. 53, are acute-, right-, or obtuse-angled, and in column B whether they are equilateral, isosceles, or scalene.

Triangle numbered	A	B
1		
2		
3		

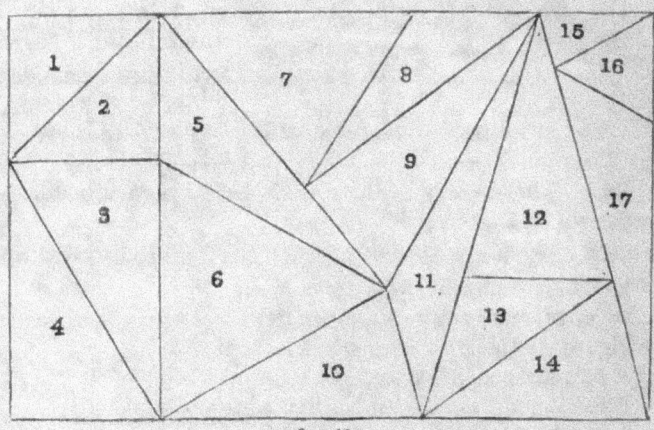

fig. 53.

¶Ex. 220. Draw a triangle ABC; produce BC to D. Measure angles A, B; what is their sum? Measure ∠ BCD. What do you infer?

¶Ex. 221. Test your inference from Ex. 220 for several different triangles.

FACT F_2. [Theorem I. 8. Cor. 1.]

If one side of a triangle is produced, the exterior angle so formed is equal to the sum of the two interior opposite angles.

†Ex. 222. Prove Fact F_2 using the construction of Ex. 197.

¶Ex. 223. In fig. 31 which are the interior angles opposite to (i) ∠ABD, (ii) ∠ACE? In fig. 32 which are the exterior angles opposite to (i) ∠BCY, (ii) ∠CBX?

ANGLES OF A POLYGON.

DEF. A plane figure bounded by straight lines is called a polygon.

DEF. A figure bounded by equal straight lines, which has all its angles equal, is called a **regular polygon.**

DEF. A figure of 3 sides is called a **triangle** (△).

,, ,, 4 ,, ,, ,, quadrilateral (4-gon).
,, ,, 5 ,, ,, ,, pentagon (5-gon).
,, ,, 6 ,, ,, ,, hexagon (6-gon).
,, ,, 7 ,, ,, ,, heptagon (7-gon).
,, ,, 8 ,, ,, ,, octagon (8-gon).

DEF. The corners of a triangle or polygon are called its **vertices.**

DEF. The sum of the sides of a polygon is called its **perimeter.**

¶Ex. 224. A yacht sails from A round the pentagon BCDEF back to A (see fig. 54). Sketch the figure freehand, and mark the angle the yacht turns through at B, C, D, E, F.

When it gets back to A, it has headed towards every point of the compass; what then is the sum of the angles through which it has turned?

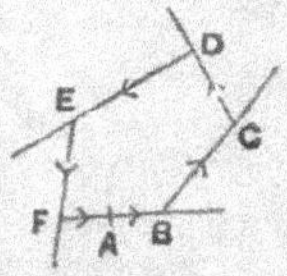

fig. 54.

¶Ex. 225. Draw a figure to show which angles a yacht turns through in sailing round a 6-sided course. What is the sum of these angles?

¶Ex. 226. Repeat Ex. 225 for a triangular course.

¶Ex. 227. Produce the sides of a square* in order (see fig. 55); what is the sum of the exterior angles?

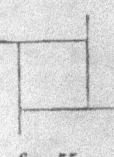

¶Ex. 228. If two of the angles of a triangle are 67° and 79°, what is the third angle? What are the exterior angles, formed by producing the sides in order? What is their sum?

fig. 55.

¶Ex. 229. What fact about the exterior angles of a polygon do you infer from the above examples?

FACT G.　[Theorem I. 9.]

If the sides of a convex† polygon are produced in order, the sum of the angles so formed is equal to four right angles.

†Ex. 230. Use Fact G to prove Fact F_1.

¶Ex. 231. What is the size of each exterior angle of a regular octagon (8-gon)? Hence find the size of each interior angle.

Ex. 232. What are the exterior angles of regular polygons of 12, 10, 5, 3 sides?

Hence find the interior angles of these polygons.

Ex. 233. The exterior angle of a regular polygon is 60°; how many sides has the polygon?

Ex. 234. How many sides have the regular polygons whose exterior angles are (i) 10°, (ii) 1°, (iii) 2½°?

* A square may be assumed to have (i) all its sides equal and (ii) all its angles right angles.

† The polygon in fig. 54 is convex; whereas such a polygon as that of fig. 56 is called **re-entrant**.

fig. 56.

Ex. 235. Is it possible to have regular polygons whose exterior angles are (i) 15°, (ii) 7°, (iii) 11°, (iv) 6°, (v) 5°, (vi) 4°?

¶**Ex. 236.** Is it possible to have regular polygons whose exterior angles are obtuse?

Ex. 237. Is it possible to have regular polygons whose interior angles are (i) 108°, (ii) 120°, (iii) 130°, (iv) 144°, (v) 60°? (Think of the exterior angles.)

In the cases which are possible, find the number of sides.

Ex. 238. Make a table showing the exterior and interior angles of regular polygons of 3, 4, 5...10 sides.

Draw a graph showing horizontally the number of sides and vertically the number of degrees in the angles.

Ex. 239. Construct a regular pentagon having each side 2 in. long.

[Calculate its angles, draw AB=2 in., at B make ∠ ABC=the angle of the regular pentagon, cut off BC=2 in., &c., &c.]

Ex. 240. Construct a regular octagon having each side 2 in. long.

Ex. 241. Construct a regular 12-gon having each side 1·5 in. long.

Ex. 242. Describe a circle of radius 5 cm. (see fig. 57); at its centre O draw two lines at right angles to cut the circle at A, B, C, D. Join AB, BC, CD, DA. Measure each of these lines and each of the angles ABC, BCD, CDA, DAB. Is the figure ABCD regular?

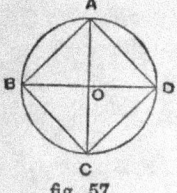

fig. 57.

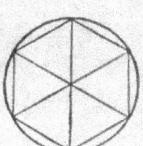

fig. 58.

Ex. 243. Describe a circle of radius 5 cm. (see fig. 58); at its centre make a set of angles each equal to $60°\left(\text{i.e. }\dfrac{360°}{6}\right)$; join the points where the arms cut the circle. Is the resulting figure regular?

Ex. 244. Calculate the angle at the centre for each of the following regular polygons;

(i) 5-gon. (ii) 8-gon, (iii) 9-gon, (iv) triangle, (v) 10-gon, (vi) 16-gon.

Ex. 245. Make a table of the results of Exx. 242-244.

Regular Polygons	
Number of sides	Angle at centre
3	120°
4	90°
5	

More Exact Measurement by Estimation.

So far you have only measured to one place of decimals in inches or centimetres, but you will often need to measure more accurately. To do this you must *imagine* each tenth of an inch (or centimetre) divided again into 10 equal parts.

The line AB is more than 1·2 in. and less than 1·3 in. ;

fig. 59.

if its length is almost exactly half-way between these measurements you will say it is 1·25 in. ;

if it is a little more than half-way you will say it is 1·26 in. ;

if it is about a third of the way you will say it is 1·23 in. ;

if it is about two-thirds of the way you will say it is 1 27 in. and so on.

With a little practice you ought to get this figure nearly accurate.

In the same way you can measure angles to within less than a degree.

¶Ex. 246. (i) What fraction of an inch does a figure in the second place of decimals represent?

(ii) What fraction of an inch is ·03?

CONSTRUCTION OF TRIANGLES FROM SUFFICIENT DATA.

A triangle has three sides and three angles; these sides and angles may be called the six parts of the triangle.

Ex. 247. How many parts has a quadrilateral?

To copy a straight-line figure, how many parts must be copied?

In copying a triangle, you might expect that all 6 parts would have to be copied, but you will find that this would be waste of labour.

Suppose that you are making a copy of a triangle ABC. Put a piece of tracing paper over the triangle and trace freehand first AB, then ∠ ABC, then BC. Then take the tracing paper away.

You can now complete the copy without tracing any more parts by joining BC.

You have thus made a copy of the triangle by copying *three only* of its parts. *The other three parts take care of themselves.*

If you are copying with tracing paper, no doubt it is as easy to copy 6 parts as to copy 3. But if you have no tracing paper, and have to copy the sides and angles with ruler and protractor, to copy 6 parts would be a waste of time.

¶Ex. 248. With ruler and protractor, copy △PQR in fig. 60 (on p. 42) by copying 3 parts only.

¶Ex. 249. Try how many parts of a quadrilateral you must copy in order to copy a quadrilateral. (Either use tracing paper, or else copy the sides and angles freehand*.)

¶Ex. 250. Repeat Ex. 249 for a 5-sided figure (pentagon).

You have just seen that you can *fix* a triangle PQR by copying PQ, ∠ PQR and QR—*two sides and the angle included (or contained) by those sides.* You have now to make up your mind if any other three parts would serve equally well to fix the copy.

* In copying sides freehand, measure the lengths roughly along your pencil; copy angles by eye; mark on your figure the parts you have copied.

¶Ex. 251. Would *any* two sides and the included angle serve to fix the copy of a triangle?

¶Ex. 252. Try if you can copy △PQR freehand* by copying two sides PQ and PR, and *an angle not included* say ∠Q. You will find that there is a difficulty here. Remember that this case has not been dealt with properly; we shall return to it later.

¶Ex. 253. Try if you can copy a triangle (drawing freehand*) by copying a side and the two angles at the ends of the side.

¶Ex. 254. With ruler and protractor, make an accurate copy of △PQR in fig. 60 by copying QR, ∠Q, and ∠R.

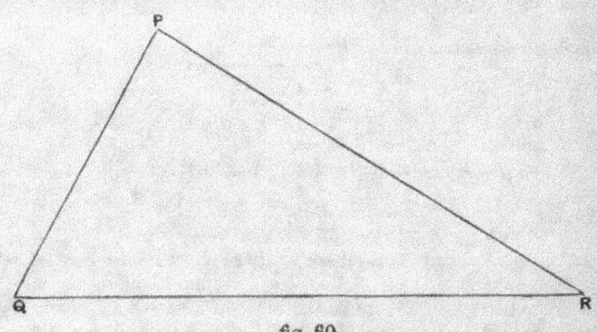

fig. 60.

¶Ex. 255. Try to construct a triangle given one side and two angles *one of which is not at the end of the given side.* Say QR = 2 in., ∠Q = 60°, ∠P = 70°. If you fail at first, think what must be the size of ∠R. Measure PQ when you have completed the copy.

Ex. 256. Construct a triangle given PR = 3 in., ∠P = 40°, ∠Q = 80°. Measure PQ.

¶Ex. 257. Try if *the lengths of three sides* are enough to fix a triangle, nothing being said about the angles. For this experi-

* See note on p. 41.

ment, take three sticks. See if you can fit them together in various ways so as to form different triangles, or if you always get the same triangle however you may arrange them.

¶Ex. 258. In the same way, try if the *lengths of four sides* are enough to fix a quadrilateral.

¶Ex. 259. Try if you can copy a triangle by copying the three angles freehand*, taking no notice of the sides. See if you are always bound to get the same triangle, or if you can get different triangles.

By this time you ought to be aware of the following facts :—

A triangle is fixed if there are given, either

     (1)   3 sides, or

     (2)   2 sides and the included angle, or

     (3)   1 side, and any two angles.

A triangle is not fixed if there are given 3 angles.

The case of 2 sides given and an angle not included has not been settled yet.

When told to construct a figure to given measurements, first make a fair-sized freehand sketch and write the given measurements on the sketch. The sketch need not be accurate, but straight lines should look straight, and angles should be drawn acute, right, or obtuse according to the given measurements.

Construction of triangles given two sides and the included angle.

¶Ex. 260‡. Make an angle ABC = 74°; cut off from its arms BC = 3·2 in., BA = 2·8 in.; join AC. Measure the remaining side and angles of the triangle ABC.

---

* See note on p. 41.

‡ In all cases measure the remaining sides (in inches or centimetres according to what units are used in the data); also measure the remaining angles.

Ex. 261*. Construct triangles to the following measurements:—

(i)   ∠ABC = 80°,   AB = 2·2 in.,   BC = 2·9 in.

(ii)§   ∠B = 28°,   AB = 7·3 cm.,   BC = 12·1 cm.

(iii)   ∠A = 42°,   AB = 3·7 in.,   CA = 3·7 in.

(iv)§   ∠B = 126°,   AB = 6·1 cm.,   BC = 6·1 cm.

(v)   ∠C = 90°,   BC = 3·9 in.,   CA = 2·8 in.

(vi)   BC = 6·7 cm.,   ∠C = 48°,   CA = 9·0 cm.

(vii)   AB = 4·7 in.,   BC = 2·9 in.,   ∠B = 32°.

(viii)§ CA = 2·6 in.,   AB = 3·3 in.,   ∠A = 162°.

(ix)   ∠C = 79°,   CA = 4·7 cm.,   BC = 6·1 cm.

(x)   AB = 4·6 cm.,   CA = 8·7 cm.,   ∠A = 58°.

Construction of triangles given one side and any two angles.

¶**Ex. 262*.** Draw a straight line AB 9 cm. long, at A make an angle BAC = 60°, at B make an angle ABC = 40°, produce AC, BC to cut at C. Measure the remaining sides and angle of the triangle ABC. What is the sum of the three angles?

Ex. 263*. Construct triangles to the following measurements:—

(In case the construction is impossible with the given measurements, try to explain why it is impossible.)

(i)   AB = 8·3 cm.,   ∠A = 45°,   ∠B = 72°.

(ii)§   AB = 3·9 in.,   ∠A = 39°,   ∠B = 39°.

(iii)   ∠B = 90°,   BC = 7·2 cm.,   ∠C = 42°.

(iv)§   ∠C = 116°,   CA = 1·8 in.,   ∠A = 78°.

(v)   ∠A = 60°,   ∠C = 60°,   AC = 6·5 cm.

(vi)   ∠B = 33°,   ∠C = 113°,   BC = 6·9 cm.

(vii)   ∠A = 73°,   ∠B = 24°,   AB = 3·2 in.

(viii)   CA = 9·2 cm.,   ∠C = 31°,   ∠A = 59°.

(ix)§   AB = 2·8 in.,   ∠A = 50°,   ∠B = 130°.

(x)   AB = 12·1 cm.,   ∠A = 27°,   ∠B = 37°.

---

* In all cases measure the remaining sides (in inches or centimetres according to what units are used in the data); also measure the remaining angles.

§ These will be enough exercises of this type unless much practice is needed.

¶Ex. 264*. Construct a triangle ABC, having ∠ A = 76°, ∠ B = 54°, BC = 2·8 in. What is ∠ C?

First find ∠C by calculation, then construct the triangle as though BC, ∠B and ∠C were given.

Measure ∠A; this will be a means of testing the accuracy of your drawing.

Ex. 265*. Construct triangles to the following measurements:—

(i)   BC = 8·0 cm.,   ∠A = 77°,   ∠B = 46°.

(ii)§   AB = 7·3 cm.,   ∠B = ∠C = 57°.

(iii)   ∠B = 114°,   ∠C = 33°,   AC = 9·4 cm.

(iv)§   ∠C = ∠A = 60°,   AB = 2·7 in.

(v)   AB = 4·3 cm.,   ∠A = 57°,   ∠C = 33°.

(vi)§   BC = 1·1 in.,   ∠A = 14°,   ∠C = 52°.

Construction of triangles given three sides.

¶Ex. 266. Take a point O on your paper and mark a number of points each of which is 2 in. from O. [To do this most easily, open your compasses 2 in., place one point at O, and mark points with the other.] The pattern you obtain is a circle; all the points 2 in. from O are on this circle.

¶Ex. 267. How does a gardener mark out a circular bed?

¶Ex. 268. Mark two points A, B, 3 in. apart.

(i) On what curve do all the points lie which are 2·7 in. from A?

(ii) On what curve do all the points lie which are 2·2 in. from B?

(iii) Is there a point which is 2·7 in. from A and also 2·2 in. from B?

(iv) Is there more than one such point?

* In all cases measure the remaining sides (in inches or centimetres according to what units are used in the data); also measure the remaining angles.

§ These will be enough exercises of this type unless much practice is needed.

Ex. 269. A and B are two points 7·4 cm. apart; find, as in Ex. 268, a point which is 5·7 cm. from A and 3·5 cm. from B.

Ex. 270. Repeat Ex. 269, without drawing the whole circles. See fig. 61.

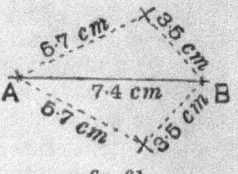

fig. 61.

¶Ex. 271. (i) Construct a triangle, the lengths of whose sides are 12·1 cm., 8·2 cm., 6·1 cm.

(ii) In how many points do your construction circles intersect?

(iii) How many triangles can you construct with their sides of the given lengths? Are these triangles **congruent** (i.e. could they be made to fit on one another exactly)?

Ex. 272* Construct triangles to the following measurements :—

(It is best to draw the longest side first.)

(i)§   BC = 8·9 cm.,   CA = 8·3 cm.,   AB = 6·7 cm

(ii)   BC = 6·9 cm.,   CA = 11·4 cm.,   AB = 5·8 cm.

(iii)§   BC = 5·3 cm.,   CA = 8·3 cm.,   AB = 2·5 cm.

(iv)   BC = 3·9 in.,   CA = 2·5 in.,   AB = 2·5 in.

(v)   BC = 3·2 in.,   CA = 3·2 in.,   AB = 1·8 in.

(vi)   BC = 6·6 cm.,   CA = 6·6 cm.,   AB = 9·3 cm.

(vii)   BC = 6·9 cm.,   CA = 6·9 cm.,   AB = 6·9 cm.

(viii)   BC = 6·5 cm.,   CA = 9·6 cm.,   AB = 7·2 cm.

(ix)§   BC = 2·1 in.,   CA = 1·1 in.,   AB = 3·2 in.

(x)§   BC = 4·1 in.,   CA = 4·1 in.,   AB = 4·1 in.

* In all cases measure the remaining sides (in inches or centimetres according to what units are used in the data); also measure the remaining angles.

§ These will be enough exercises of this type unless much practice is needed.

Construction of quadrilaterals, etc., from sufficient data.

Ex. 273*. Construct quadrilaterals ABCD to the following measurements :—

(Here it is especially important that, before beginning the construction, a rough sketch should be made showing the given parts.

Note that the letters must be taken in order round the quadrilateral; e.g. the quadrilateral in fig. 62 is called ABCD and *not* ABDC.)

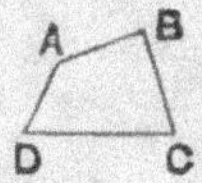

fig. 62.

  (i) AB = 6·3 cm., ∠B = 82°, BC = 8·2 cm., ∠C = 90°, CD = 7·7 cm.

  (ii) AB = 3·4 in., BC = 2·2 in., AD = 2·9 in., ∠A = 68°, ∠B = 86°.

  (iii) ∠B = 116°, BC = 1·4 in., ∠C = 99°, CD = 1·9 in., ∠D = 92°.

  (iv) ∠A = 67°, ∠B = 113°, ∠D = 46°, AB = 5·3 cm., AD = 8·6 cm.

  (v) ∠B = 122°, ∠C = 130°, ∠D = 130°, BC = CD = 1·6 in.

  (vi) AD = 3·0 in., ∠D = 118°, ∠DAC = 27°, ∠BAC = 35°, AB = 2·4 in.

  (vii) AC = 5·6 cm., ∠BAC = 58°, ∠DAC = 69°, ∠BCA = 58°, ∠DCA = 69°.

  (viii) AB = 1·9 in., BD = 1·7 in., CD = 2·0 in., ∠ABD = 118°, ∠BDC = 23°.

Ex. 274*. Construct quadrilaterals ABCD to the following measurements :—

  (i) AB = 2·3 in., BC = 2·1 in., CD = 3·3 in., DA = 1·5 in., BD = 3·4 in.

  (ii) AB = CD = 6·4 cm., BC = DA = 3·7 cm., BD = 5·7 cm.

  (iii) AB = AD = 1·9 in., CB = CD = 2·9 in., BD = 2·5 in.

  (iv) AB = BC = CD = DA = 5·1 cm., AC = 9·2 cm.

  (v) AB = 3·8 in., BC = 1·7 in., CD = 1·0 in., DA = 4·9 in., ∠B = 146°.

  (vi) AB = 5·3 cm., BC = 6·3 cm., CD = 6·7 cm., ∠B = 70°, ∠C = 48°.

* See note on p. 44.

(vii) AB = 2·7 cm., BC = 7·5 cm., AD = 8·4 cm., ∠C = 98°, ∠ DBC = 28°.

(viii) BC = CD = 2·4 in., BD = 1·9 in., ∠ ABD = ∠ ADB = 67°.

(ix) AB = 9·3 cm., BC = DA = 6·7 cm., ∠A = 111°, ∠B = 28°.

Ex. 275*. Construct pentagons ABCDE to the following measurements :—

(i) AB = 2·0 in., BC = 2·2 in., CD = 1·7 in., DE = 2·2 in., EA = 2·5 in., ∠B = 111°, ∠C = 149°.

(ii) AB = 1·7 in., BC = 1·0 in., CD = 2·2 in., DE = 3·4 in., EA = 0·5 in., ∠A = 126°, ∠B = 137°.

(iii) AB = 5 cm., BC = 3·7 cm., CD = 3·6 cm., DE = 4·3 cm., EA = 3·8 cm., AC = 6·4 cm., AD = 6·7 cm.

(iv) AB = BC = CD = DE = EA = 5·0 cm., AC = BE = 8·1 cm.

Construction of triangles—miscellaneous data.

Ex. 276*. Construct triangles to the following measurements :—

(i)	BC = 3·18 in.,	AB = 3·18 in.,	∠B = 33·5°.
(ii)	BC = 2·39 in.,	CA = 2·44 in.,	∠C = 63·5°.
(iii)	AB = 2·82 in.,	AC = 2·77 in.,	∠A = 137°.
(iv)	AB = 3·00 in.,	∠A = 61°,	∠B = 59°.
(v)	BC = 3·52 in.,	∠B = 25°,	∠C = 23°.
(vi)	AC = 10·65 cm.,	∠A = 54·5°,	∠C = 36°.
(vii)	BC = 6·40 cm.,	CA = 9·05 cm.,	AB = 7·63 cm.
(viii)	BC = 7·69 cm.,	CA = 9·30 cm.,	AB = 5·30 cm.
(ix)	BC = 4·53 in.,	CA = 2·68 in.,	AB = 2·02 in.
(x)	AB = 2·71 in.,	∠B = 55·5°,	∠C = 67·5°.
(xi)	∠A = 24°,	∠C = 47·5°,	BC = 3·04 cm.
(xii)	∠A = 133°,	BC = 10·73 cm.,	∠B = 23·5°.
(xiii)	∠C = 90°,	BC = 1·00 in.,	CA = 2·00 in.
(xiv)	BC = 4·09 cm.,	CA = 3·31 cm.,	AB = 7·54 cm.
(xv)	∠A = 90·5°,	∠B = 78°,	BC = 3·54 in.
(xvi)	AB = 2·99 in.,	∠B = 127·5°,	∠C = 53·5°.
(xvii)	AB = 2·92 in.,	∠B = 59°,	AC = 2·39 in.
(xviii)	∠B = 33·5°,	BC = 2·61 in.,	CA = 1·54 in.
(xix)	CB = 2·16 in.,	CA = 2·64 in.,	∠B = 64·5°.
(xx)	∠A = 24°,	AB = 7·76 cm.,	BC = 2·87 cm.

* See note on p. 44.

CONGRUENT TRIANGLES.

If two figures when applied to one another can be made to **coincide** (i.e. fit exactly) they must be equal in all respects.

Figures which are equal in all respects are said to be **congruent.**

The sign ≡ is used to denote that figures are congruent.

You have learnt that if you copy two sides and the contained (or included) angle of a triangle, you make a triangle congruent with the first. This fact may be stated as follows :—

FACT H. [Theorem I. 10.]

If two triangles have two sides of the one equal to two sides of the other, each to each, and also the angles contained by those sides equal, the triangles are congruent.

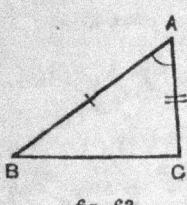
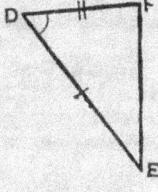

fig. 63. fig. 64.

N.B. *It must be noted carefully that the congruence of the triangles cannot be inferred unless the equal angles are the angles contained by the sides which are given equal.*

Ex. **277.** Make a list of all the equal sides and angles in △' ABC and DEF of figs. 63, 64. Which were given equal?

†Ex. **278.** Draw two triangles PQR, XYZ and make QR = XY, RP = YZ, and ∠Q = ∠Z. Would this prove the triangles congruent? Give two reasons.

†Ex. **279***. ABCD is a square, E is the mid-point of AB; equal lengths AP and BQ are cut off from AD and BC. Join EP and EQ. Prove that △ AEP ≡ △ BEQ.

Write down all the pairs of lines and angles in these triangles which you have proved equal.

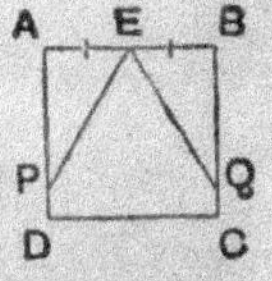

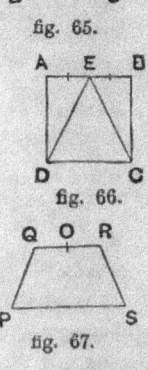

fig. 65.

†Ex. **280***. ABCD is a square, E is the mid-point of AB; join CE and DE. Prove that △ AED ≡ △ BEC.

Write down all the pairs of lines and angles in these triangles which you have proved equal.

fig. 66.

†Ex. **281**. PQRS is a quadrilateral in which PQ=SR, ∠Q=∠R, and O is the mid-point of QR. Prove that OP=OS.

[You must first join OP and OS, and mark in your figure all the parts that are given equal; you will then see that you want to prove that △OQP ≡ △ORS.]

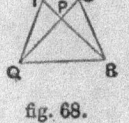

fig. 67.

†Ex. **282***. ABCD is a square; E, F, G are the mid-points of AB, BC, CD respectively. Join EF and FG and prove them equal.

[Which are the two triangles that you must prove equal?]

†Ex. **283**. ABC, DEF are two triangles which are equal in all respects; X is the mid-point of BC, Y is the mid-point of EF. Prove that AX=DY, and ∠AXB=∠DYE.

[You will of course have to join AX and DY.]

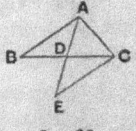

†Ex. **284**. The equal sides QP, RP of an isosceles triangle PQR are produced to S, T so that PS=PT; prove that TQ=SR.

fig. 68.

†Ex. **285**. D is the mid-point of the side BC of a △ ABC, AD is produced to E so that DE=AD. Prove that AB=EC and that AB, EC are parallel.

[First prove a pair of triangles congruent.]

fig. 69.

* The following properties of a square may be assumed (i) all the sides are equal, (ii) all the angles are right angles.

†Ex. **286.** Show that the distance between G and H, the opposite corners of a house, can be found as follows. At a point P set up a post; step off HP and an equal distance PN, taking care to keep in a straight line with the post and the corner H; step off GP and an equal distance PM, M being in the same straight line as G and P. Measure MN; this must be equal to GH.

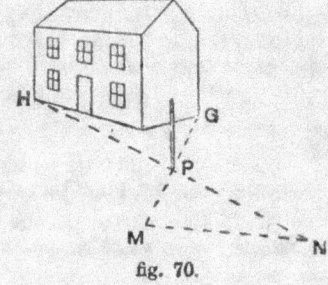

fig. 70.

Draw a ground plan and prove that MN=GH.

†Ex. **287.** W is the mid-point of a straight line YZ, WX is drawn at right angles to YZ. Prove that XY=XZ.

[A line which is a side of each of two triangles is said to be **common** to the two triangles.]

fig. 71.

You have learnt that if you copy two angles and a side (not necessarily the side between the angles) of a triangle, you make a triangle congruent with the first. This fact may be stated as follows :—

FACT J. [Theorem I. 11.]

If two triangles have two angles of the one equal to two angles of the other, each to each, and also one side of the one equal to the corresponding side of the other, the triangles are congruent.

†Ex. **288.** Draw two △ˢ GHK, XYZ, and mark GH=XY, ∠H=∠Y, and ∠K=∠X; are the triangles congruent?

†Ex. **289.** ABCD is a square, E is the mid-point of AB; at E make ∠AEP=60° and ∠BEQ=60°; let EP, EQ cut AD, BC at P and Q respectively. Prove that AP=BQ. (See fig. 65.)

†Ex. **290.** In a △XYZ, ∠Y=∠Z; XW is drawn so that ∠X is bisected; prove that XY=XZ. (See fig. 71.)

4—2

†Ex. 291. If the bisector of an angle of a triangle cuts the opposite side at right angles, the triangle must be isosceles.

[Let XYZ be a triangle; and let XW, the bisector of ∠X, cut YZ at right angles at W; prove that XY=XZ. See fig. 71.]

†Ex. 292. ABC, DEF are two triangles which are equal in all respects: AP, DQ are drawn perpendicular to BC, EF respectively. Prove that AP=DQ.

†Ex. 293. △ABC≡△DEF. AG, DH are the bisectors of ∠A, ∠D and meet the opposite sides in G, H. Prove that AG=DH.

†Ex. 294. The following method may be used to find the breadth of a river. Choose a place where the river is straight, note some conspicuous object T (e.g. a tree) on the edge of the other bank; from a point O opposite T measure a distance OS along the bank; put a stick in the ground at S; walk on to a point P such that SP=OS; from P walk at right angles to the

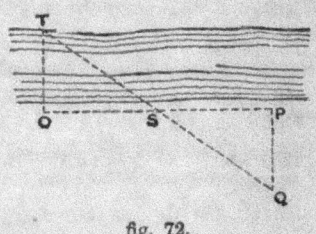

fig. 72.

river till you are in the same straight line as S and T. PQ is equal to the breadth of the river. Prove this.

You have learnt that if you copy three sides of a triangle, you make a triangle congruent with the first. This fact may be stated as follows:—

Fact K. [Theorem i. 14.]

If two triangles have the three sides of the one equal to the three sides of the other, each to each, the triangles are congruent.

†Ex. 295. State the converse of this theorem. Is it true?

†Ex. 296. If, in a quadrilateral ABCD, AB=AD, CB=CD, prove that AC bisects ∠A and ∠C.

†Ex. **297.** PQ and RS are two equal chords of a circle whose centre is O. Prove that ∠POQ=∠ROS.

(A **chord** of a circle is a straight line joining any two points on the circle.)

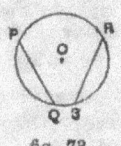

†Ex. **298.** AB is a chord of a circle whose centre is O; C is the mid-point of the chord AB. Show that OC is perpendicular to AB.

fig. 73.

†Ex. **299.** If the opposite sides of a quadrilateral are equal, it is a parallelogram.

[Draw a diagonal.]

†Ex. **300.** The bisector of the angle between the equal sides of an isosceles triangle is perpendicular to the base.

[Let **XYZ** be an isosceles triangle, having **XY=XZ**; let **XW** bisect ∠**YXZ** and let it meet **YZ** at **W**; prove ∠**XWY**=∠**XWZ**. See fig. 71.]

†Ex. **301.** XYZ is an isosceles triangle having **XY=XZ**; prove that ∠Y=∠Z.

[Draw **XW** the bisector of ∠**YXZ**.]

†Ex. **302.** The perpendiculars drawn to the arms of an angle from any point on the bisector of the angle are equal to one another.

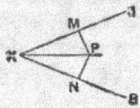

†Ex. **303.** ABCD is a parallelogram; prove that AB=CD.

[Join AC.]

fig. 74.

†Ex. **304.** Equal lengths AB, AC are cut off from the arms of an angle BAC; on BC a triangle BCD is drawn having BD=CD. Show that AD bisects ∠BAC.

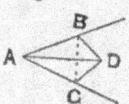

fig. 75.

†Ex. **305.** OA, OB, OC are three radii of a circle. If ∠AOB=∠COB, prove that BO bisects AC.

†Ex. **306.** If the diagonal PR of a quadrilateral PQRS bisects the angles at P and R, prove that the quadrilateral has two pairs of equal sides.

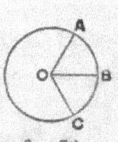

†Ex. **307.** In a quadrilateral ABCD, AD=BC and the diagonals AC, BD are equal; prove that ∠ADC=∠BCD.

fig. 76.

Also prove that, if AC, BD intersect at O, △OCD is isosceles.

†Ex. 308. In fig. 77, AB and DC are equal and
parallel; prove that AD=BC.

[Join BD. Since AB is parallel to DC ∴ ∠?=∠?.]

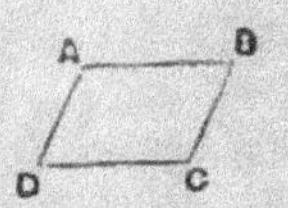

†Ex. 309. The equal sides AB, AC of an isosceles
triangle ABC are produced to X and Y respectively;
BX is made equal to CY (see fig. 82). If ∠CBX=∠BCY, prove that
CX=BY.

fig. 77.

[By the side of your figure make sketches of the triangles BCX, CBY.]

†Ex. 310. XY is a straight line, XP and YQ are
drawn at right angles to XY, and XP is made equal to
YQ. Prove that ∠PYX=∠QXY.

†Ex. 311. A triangle XYZ has ∠Y=∠Z; prove that
the perpendiculars from the mid-point of YZ to XY and
XZ are equal to one another.

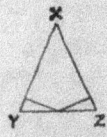

fig. 78.

†Ex. 312. A triangle ABC has ∠B=∠C; prove that
the perpendiculars from B and C on the opposite sides
are equal to one another.

†Ex. 313. A triangle ABC has AB=AC; prove that
the perpendiculars from B and C on the opposite sides
are equal to one another.

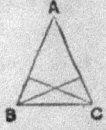

†Ex. 314. ABCD is a quadrilateral in which AB=CD,
AD=BC and ∠A=∠C; prove that ABCD is a parallelogram.

fig. 79.

[Join BD.]

†Ex. 315. If the diagonals of a quadrilateral bisect one another it must
be a parallelogram.

†Ex. 316. The diagonal AC of a quadrilateral ABCD bisects the angle
A and ∠ABC=∠ADC; does BC=CD?

¶Ex. 317. Draw two or three isosceles triangles; measure
their angles. What fact do you notice?

†Ex. 318. Draw an isosceles triangle PQR, having PQ=PR.
Let the bisector of ∠QPR meet QR in X. Prove that △ˢ PQX,
PRX are congruent. What do you deduce from this as to the
angles of an isosceles triangle?

FACT L. [Theorem I. 12.]

If two sides of a triangle are equal, the angles opposite to these sides are equal.

The phrase "the sides" of an isosceles triangle is often used to mean the equal sides, "the base" to mean the other side, "the vertex" to mean the point at which the equal sides meet, and "the vertical angle" to mean the angle at the vertex.

Ex. **319.** In a triangle XYZ, XY = XZ; find the angles of the triangle in the following cases: (i) ∠Y = 74°, (ii) ∠X = 36°, (iii) ∠X = 142°, (iv) ∠Y = 13°, (v) ∠Z = 97°, (vi) ∠Z = 45°.

†Ex. **320.** Each base angle of an isosceles triangle must be acute.

Ex. **321.** Find the angles of an isosceles triangle in which each of the base angles is half of the vertical angle.

Ex. **322.** Find the angles of an isosceles triangle in which each of the base angles is double of the vertical angle.

†Ex. **323. Prove that a triangle which is equilateral is also equiangular.** (See definition, p. 36.)

[If PQR is an equilateral triangle, ∵ QP = QR ∴ ∠? = ∠?.]

Ex. **324.** In a triangle ABC, AB = 9·2 cm., ∠C = 82°, AC = 9·2 cm.; AB, AC are produced to D, E respectively. Find all the angles in the figure.

†Ex. **325.** ABC is an isosceles triangle; the equal sides AB, AC are produced to X, Y respectively. Prove that ∠XBC = ∠YCB.

†Ex. **326.** EDA, FDA are two isosceles triangles on opposite sides of the same base DA; prove that ∠EDF = ∠EAF.

†Ex. **327.** EDA, FDA are two isosceles triangles on the same side of the same base DA; prove that ∠EDF = ∠EAF.

†Ex. **328.** Through the vertex P of an isosceles triangle PQR a straight line XPY is drawn parallel to QR; prove that ∠QPX = ∠RPY.

¶Ex. 329. Draw a line BC; at B and C make equal angles CBA, BCA so as to form a triangle ABC. Measure AB and AC.

¶Ex. 330. Repeat Ex. 329 two or three times with other lines and angles.

¶Ex. 331. What fact do you infer from Exx. 329, 330?

†Ex. 332. In △XYZ, ∠Y is given equal to ∠Z. Let XN, the bisector of ∠YXZ, meet YZ in N. Prove △ˢ YXN, ZXN congruent. What does this show about the sides of △XYZ.

Fact M.   [Theorem i. 13.]

If two angles of a triangle are equal, the sides opposite to these angles are equal.

†Ex. **333. Prove that if a triangle PQR is equiangular, it must also be equilateral.**

[∠Q=∠R, ∴ side?=side?.]

†Ex. **334.** The sides AB, AC of a triangle are produced to XY; prove that, if ∠XBC=∠YCB, △ABC is isosceles. (See fig. 32.)

†Ex. **335.** A straight line drawn parallel to the base of an isosceles triangle to cut the equal sides forms another isosceles triangle. (See fig. 80.)

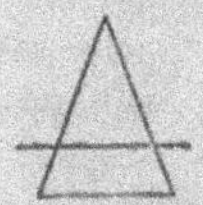

fig. 80.

†Ex. **336.** XYZ is an isosceles triangle; the bisectors of the equal angles (Y, Z) meet at O; prove that △OYZ is also isosceles. (See fig. 88.)

†Ex. **337.** From Q and R, the extremities of the base of an isosceles triangle PQR, perpendiculars are drawn to the opposite sides. If these perpendiculars intersect at X, prove that XQ=XR.

†Ex. **338.** XYZ is an isosceles triangle (XY=XZ); the bisectors of ∠Y and ∠Z meet at O; prove that OX bisects ∠X.

†Ex. **339.** From the mid-point O of a straight line AB a straight line OC is drawn; if OC=OA, ∠ACB is a right angle.

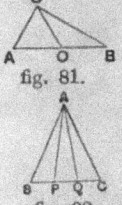

fig. 81.

†Ex. **340.** In fig. 82, △ABC is isosceles and BP=CQ; prove that ∠APQ=∠AQP.

[First prove AP=AQ.]

†Ex. **341.** The perpendicular from the vertex of an isosceles triangle to the base bisects the base.

fig. 82.

†Ex. **342.** If through any point in the bisector of an angle a line is drawn parallel to either of the arms of the angle, the triangle thus formed is isosceles.

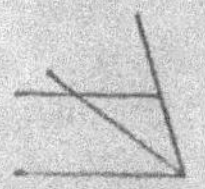

†Ex. **343.** ABCD is a quadrilateral in which AB=AD, and ∠B=∠D; prove that CB=CD.

[Draw a diagonal.]

fig. 83.

†Ex. **344.** The straight lines joining the mid-point of the base of an isosceles triangle to the mid-points of the sides are equal. (See fig. 84.)

fig. 84.

†Ex. **345.** If A, B are the mid-points of the equal sides XY, XZ of an isosceles triangle, prove that AZ=BY.

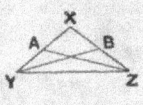

†Ex. **346.** The bisectors of the equal angles Y, Z of an isosceles triangle XYZ meet at O. Prove that XO bisects ∠X.

fig. 85.

†Ex. **347.** In the base BC of a triangle ABC, points P, Q are taken such that ∠BAP=∠CAQ; if AP=AQ, prove △ABC is isosceles.

†Ex. **348.** The bisectors of the base angles of an isosceles triangle are equal.

fig. 86.

†Ex. **349.** At the ends of the base BC of an isosceles triangle ABC, perpendiculars are drawn to the base to meet the equal sides produced; prove that these perpendiculars are equal.

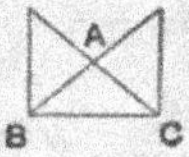

fig. 87.

†Ex. **350.** EDA, FDA are two isosceles triangles on opposite sides of the same base DA; prove that EF bisects DA at right angles.
[First prove △DEF≡△AEF.]

†Ex. **351.** In a quadrilateral ABCD, ∠ˢ A, B are equal and obtuse, and AB is parallel to CD; prove that AD=BC.

[Produce DA, CB till they meet.]

†Ex. **352.** XYZ is an isosceles triangle (XY=XZ), the bisectors of ∠X and ∠Z meet at O; prove that OY bisects ∠Y.

†Ex. **353.** The angle between a diagonal and a side of a square is 45°.

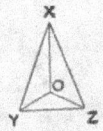

†Ex. **354.** If the ∠ˢ G, H of a triangle FGH are each   fig. 88. double of ∠F, and if the bisector of ∠G meets FH in K, prove that FK=GK=GH.

†Ex. **355.** In a quadrilateral ABCD, AD=BC and the diagonals AC, BD are equal; prove that ∠ADC=∠BCD.

Also prove that, if AC, BD intersect at O, △OCD is isosceles.

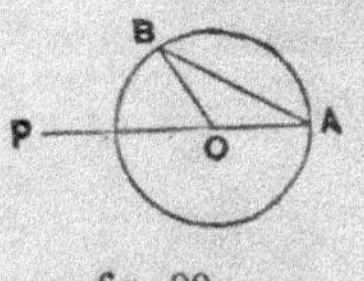

†Ex. **356.** OA, OB are radii of a circle, AO is produced to P; prove that ∠BOP=2∠BAP.

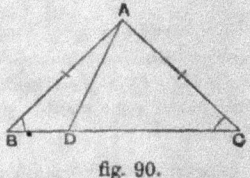

fig. 89.

†Ex. **357.** In fig. 89, prove that the perpendicular from O to AB bisects AB.

†Ex. **358.** **Two circles intersect at X, Y; prove that XY is bisected at right angles by the straight line joining the centres of the two circles.**

[Join the centres of the circles to X and Y.]

Under Fact H it was emphasized that triangles with two pairs of equal corresponding sides, and one pair of equal corresponding angles, cannot be argued congruent unless the equal angles are *contained* between the equal sides.

Look at fig. 90. Here AB = AC, and therefore ∠ C = ∠ B. Thus in △ˢ ABD, ACD we have AB = AC, AD common, and ∠ B = ∠ C. But the triangles are obviously not congruent, the equal angles not being the *contained* angles.

fig. 90.

There is one case, however, in which congruence can be argued even though the equal angles are *not* contained by the equal sides; i.e. the case in which the equal angles are right angles.

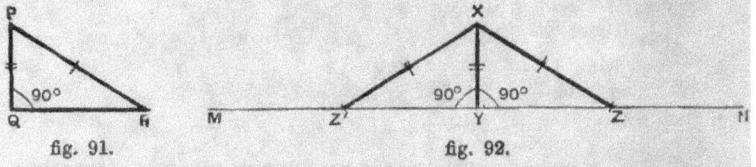

fig. 91.   fig. 92.

△PQR in fig. 91 has ∠Q a rt∠. Try to copy △PQR by

copying PQ, PR, and ∠ PQR. Make XY = PQ; then draw MYN ⊥ to XY. We have so far copied PQ and ∠ PQR. To copy PR, with centre X and radius = PR, draw a circle cutting MN in Z and Z′. We now seem to have not one copy of △ PQR, but two copies;—△ XYZ and △ XYZ′.

But *these two triangles are congruent.*

This you can prove as follows:—

(i)   Show that ∠ XZY = ∠ XZ′Y.

(ii)   Now show that △ XYZ ≡ △ XYZ′.

We therefore arrive at

FACT N.   [Theorem I. 15.]

If two right-angled triangles have their hypotenuses* equal, and one side of the one equal to one side of the other, the triangles are congruent.

NOTE.   In using this fact to prove two triangles congruent, always begin by saying:—

In the *right-angled* triangles PQR, XYZ (see figs. 91, 92) ∠ˢ PQR, XYZ are rt ∠ˢ etc. etc.

†Ex. 359.   In fig. 65, given that ABCD is a square, E the mid-point of AB, and EP = EQ, prove that △ AEP ≡ △ BEQ.

†Ex. 360.   In a triangle XYZ, XY = XZ, and XW is drawn at right angles to YZ: prove that △ XYW ≡ △ XZW.

†Ex 361.   Perpendiculars are drawn from a point P to two straight lines XA, XB which intersect at a point X, prove that, if the perpendiculars are equal, PX bisects ∠ AXB.   (See fig. 74.)

†Ex. 362.   AB is a chord of a circle whose centre is O.   Show that the perpendicular from O on AB bisects AB.

†Ex. 363.   The perpendiculars from the centre of a circle on two equal chords of the circle are equal to one another.   (See fig. 73; use Ex. 362.)

---

* The **hypotenuse** of a right-angled triangle is the side opposite to the right angle.

†Ex. **364**.   In fig. 93, PM, QN are drawn perpendicular to the diameter AOB, O being the centre of the
circle; show that, if PM = QN, then ∠POM = ∠QON.

†Ex. **365**.   If the perpendiculars from the mid-point
of the base of a triangle to the other two sides are equal,
the triangle is isosceles.   (See fig. 78.)

fig. 93.

†Ex. **366**.   If the perpendiculars from two corners of a triangle to the
opposite sides are equal, the triangle is isosceles.   (See fig. 79.)

†Ex. **367**.   From the vertices A, X of two acute-angled triangles ABC,
XYZ, lines AD, XW are drawn perpendicular to BC, YZ respectively.   If
AD = XW, AB = XY, and AC = XZ, prove that the triangles ABC, XYZ are
congruent.

†Ex. **368**.   With the same notation as in Ex. 367, prove that, if AD = XW,
AB = XY, and BC = YZ, the triangles are congruent.

MISCELLANEOUS EXERCISES.

†Ex. **369**.   How many diagonals can be drawn through one vertex of
(i) a quadrilateral, (ii) a hexagon, (iii) a n-gon?

†Ex. **370**.   **The bisectors of the four angles formed by two intersecting straight lines are two straight lines at right angles to one
another.**

†Ex. **371**.   If the bisector of an exterior angle of a triangle is parallel to
one side, the triangle is isosceles.

†Ex. **372**.   The bisectors of the angles B, C of a triangle ABC intersect
at I; prove that ∠BIC = $90° + \frac{1}{2}$∠A.

†Ex. **373**.   The internal bisectors of two angles of a triangle can never
be at right angles to one another.

†Ex. **374**.   AB, CD are two parallel straight lines drawn in the same direction, and P is any point between them.   Prove that ∠BPD = ∠ABP + ∠CDP.

†Ex. **375**.   ABC is an isosceles triangle (AB = AC).   A straight line is
drawn at right angles to the base and cuts the sides or sides produced in D
and E.   Prove that △ADE is isosceles.

†Ex. **376**.   From the extremities of the base of an isosceles triangle
straight lines are drawn perpendicular to the opposite sides; show that the
angles which they make with the base are each equal to half the vertical
angle.

†Ex. 377. The medians* of an equilateral triangle are equal.

†Ex. 378. The bisector of the angle A of a triangle ABC meets BC in D, and BC is produced to E. Prove that ∠ABC + ∠ACE = 2∠ADC.

†Ex. 379. From a point O in a straight line XY, equal straight lines OP, OQ are drawn on opposite sides of XY so that ∠YOP = ∠YOQ. Prove that △PXY ≡ △QXY.

†Ex. 380. The sides AB, AC of a triangle are bisected in D, E; and BE, CD are produced to F, G, so that EF = BE and DG = CD. Prove that FAG is a straight line. (Show that ∠FAD = ∠CBA, etc.)

†Ex. 381. If the straight lines bisecting the angles at the base of an isosceles triangle be produced to meet, show that they contain an angle equal to an exterior angle at the base of the triangle.

†Ex. 382. XYZ is an isosceles right-angled triangle (XY = XZ); YR bisects ∠Y and meets XZ at R; RN is drawn perpendicular to YZ. Prove that RN = XR.

†Ex. 383. If two of the bisectors of the angles of a triangle meet at a point I the perpendiculars from I to the sides are all equal.

†Ex. 384. The perpendicular bisectors of two sides of a triangle meet at a point which is equidistant from the vertices of the triangle.

†Ex. 385. In the equal sides PQ, PR of an isosceles triangle PQR points X, Y are taken equidistant from P; QY, RX intersect at Z. Prove that △ˢ ZQR, ZXY are isosceles.

†Ex. 386. ABC is a triangle right-angled at A; AD is drawn perpendicular to BC. Prove that the angles of the triangles ABC, DBA are respectively equal.

†Ex. 387. From a point O in a straight line XOX' two equal straight lines OP, OQ are drawn so that ∠POQ is a right angle. PM and QN are drawn perpendicular to XX'. Prove that PM = ON.

†Ex. 388. If points P, Q, R are taken in the sides AB, BC, CA of an equilateral triangle such that AP = BQ = CR, prove that PQR is equilateral.

†Ex. 389. ABC is an equilateral triangle; DBC is an isosceles triangle on the same base BC and on the same side of it, and ∠BDC = ½∠BAC. Prove that AD = BC.

* A median is a straight line joining a vertex of a triangle to the mid-point of the opposite side.

†Ex. **390.** How many sides has the polygon, the sum of whose interior angles is three times the sum of its exterior angles?

[What is the sum of all the exterior and interior angles? What is the sum of an exterior angle and the corresponding interior angle?]

†Ex. **391.** If two isosceles triangles have equal vertical angles and if the perpendiculars from the vertices to the bases are equal, the triangles are congruent.

†Ex. **392.** If, in two quadrilaterals ABCD, PQRS,

AB=PQ, BC=QR, CD=RS, ∠B = ∠Q, and ∠C = ∠R,

the quadrilaterals are congruent.

[Prove this by joining BD and QS and proving triangles congruent.]

†Ex. **393.** If two quadrilaterals have the sides of the one equal respectively to the sides of the other taken in order, and have also one angle of the one equal to the corresponding angle of the other, the quadrilaterals are congruent.

[Draw a diagonal of each quadrilateral, and prove triangles congruent.]

†Ex. **394.** If AA', BB', CC' are diameters of a circle, prove

△ABC ≡ △A'B'C'.

†Ex. **395.** On the sides AB, BC of a triangle ABC, squares ABFG, BCED are described (on the opposite sides to the triangle); prove that

△ABD ≡ △FBC.

†Ex. **396.** On the sides of any triangle ABC, equilateral triangles BCD, CAE, ABF are described (all pointing outwards); prove that AD, BE, CF are all equal. [Prove △ˢ ACD, ECB congruent.]

†Ex. **397.** The side BC of a triangle ABC is produced to D; ∠ACB is bisected by the straight line CE which cuts AB at E. A straight line is drawn through E parallel to BC, cutting AC at F and the bisector of ∠ACD at G. Prove that EF=FG.

†Ex. **398.** ABC, DBC are two congruent triangles on opposite sides of the same base BC; prove that either AD is bisected at right angles by BC, or AD and BC bisect one another.

†Ex. **399.** In a triangle ABC, the bisector of the angle A and the perpendicular bisector of BC intersect at a point D; from D, DX, DY are drawn perpendicular to the sides AB, AC produced if necessary. Prove that

AX=AY and BX=CY.

[Join BD, CD.]

PRACTICAL CONSTRUCTION OF PARALLELS AND PERPENDICULARS.

To draw a parallel to a given line QR through a given point P by means of a set square and a straight edge.

It is important that the straight edge should not be bevelled (if it is bevelled the set square will slip over it); in the figures below, a ruler with an unbevelled edge is represented, but the base of the protractor or the edge of another set square will do equally well.

Place a set square so that one of its edges lies along the given line QR (as at (i)); hold it in that position and place the straight (unbevelled) edge in contact with it; now hold the straight edge firmly and slide the set square along it. The edge which originally lay along QR will always be parallel to QR. Slide the set square till this edge passes through P (as at (ii)), hold it firmly and rule the line.

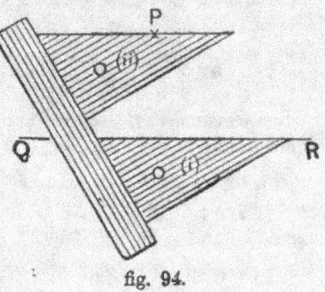

fig. 94.

Ex. 400. Draw a straight line QR and mark a point P; through P draw a parallel to QR.

Ex. 401. Repeat Ex. 400 several times using the different edges of the set square. (See fig. 95.)

Ex. 402. Near the middle of your paper draw an acute-angled triangle; through each vertex draw a line parallel to the opposite side.

To draw through a given point P a straight line perpendicular to a given straight line QR.

The difficulty of drawing a line right to the corner of a set square can be overcome as follows:—

Place a set square so that one of the edges containing the right angle lies along the given line QR (as at (i)); place the straight edge in contact with the side opposite the right angle; now hold the straight edge firmly and slide the set square

along it; the edge which lay along QR will always be parallel
to QR and the other edge
containing the right angle
will always be perpendicular to QR. Slide the set
square till this other edge
passes through P; then
draw the perpendicular.

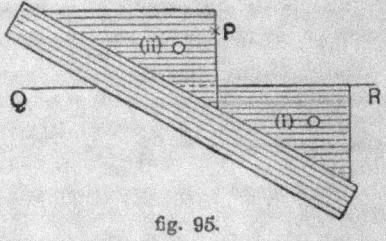

In drawing a perpen-
dicular to a given line,

fig. 95.

make the perpendicular project slightly beyond the line, so that
you may have a well-defined point of intersection.

Ex. 403. Through a given point in a straight line draw a perpendicular
to that line.

Ex. 404. Draw an acute-angled triangle; from each vertex draw a perpendicular to the opposite side.

Ex. 405. Repeat Ex. 404 with an obtuse-angled triangle. (You will
find it necessary to produce two of the sides.)

Ex. 406. Draw an acute-angled triangle; from the middle point of
each side draw a straight line at right angles to that side.

Ex. 407. Repeat Ex. 406 with an obtuse-angled triangle.

How to copy a given Rectilinear Figure.

A rectilinear figure is a figure made up of straight lines.

An exact copy of a given rectilinear figure may be made in
various ways.

1st method. Suppose that it is required to copy a pentagon
ABCDE (as in fig. 96). First copy side AB; then ∠ ABC; then
side BC; then ∠ BCD; etc. You will not find it necessary to
copy *all* the sides and angles.

Ex. 408. Draw a good-sized quadrilateral; copy it by Method I. If you
have tracing paper, make the copy on this; then see if it fits the original.

Ex. 409. Repeat Ex. 408 with an (irregular) pentagon.

2nd method. A simpler way is to prick holes through the different vertices of the given figure on to a sheet of paper below; then join the holes on the second sheet by means of straight lines.

3rd method. Place a sheet of **tracing paper** over the given figure, and mark on the tracing paper the positions of the different vertices. Then join up with straight lines.

4th method—**by intersecting arcs.**

To copy ABCDE by this method (see fig. 96). Make A′B′ = AB.

With centre A′ and radius equal to AC describe an arc of a circle.

With centre B′ and radius equal to BC describe an arc of a circle.

Let these arcs intersect at C′. Then C′ is the copy of C.

Similarly, fix D′ by means of the distances A′D′ and B′D′; fix E′ by means of the distances A′E′ and B′E′.

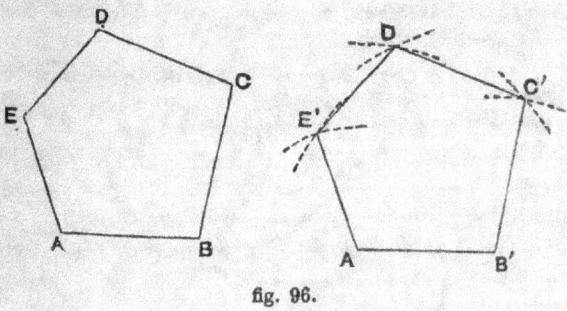

fig. 96.

The five vertices A′B′C′D′E′ are now fixed, and the copy may be completed by joining up.

In Exx. 410-412 the copies should be made on **tracing paper** if possible; the copies can then be tested by fitting on to the originals.

Ex. **410**. Draw, and copy (i) a quadrilateral, (ii) a pentagon, by the method of intersecting arcs. If tracing paper is not used, the copy may be checked by comparing its angles with those of the original.

Ex. **411**. By intersecting arcs, copy fig. 2.

Ex. **412**. By intersecting arcs, copy the part of fig. 53 which consists of triangles 1, 2, 3, 4, 5, 6.

DRAWING TO SCALE.

When drawing a map, or plan, to scale you should *always begin by making a rough sketch* showing the given dimensions, and then work from the sketch. Choose any suitable scale unless a scale is given. *And always state what scale you are using.*

Ex. 413. Draw a plan of a room 30 ft. by 22 ft.; find the distances between opposite corners. (Scale 2 ft. to 1 cm.)

Ex. 414. Draw a plan of a rectangular field 380 yards by 270 yards. What is the distance between the opposite corners?

Ex. 415. The legs of a pair of compasses are 10 cm. long. I open them to an angle of 35°. What is the distance between the compass points?

The **bearing** of a place A from a second place B is the point of the compass towards which a man at B would be facing if he were looking in the direction of A.

By "N. 10° W." or "10° W. of N." is meant the direction in which you would be looking if you first faced due north and then turned through an angle of 10° towards the west.

Ex. 416. A is 2·5 miles W. of B, and C is 4·5 miles S. of A. What is the distance from B to C? What is the bearing of B from C, and of C from B? (Scale 1 mile to 1 inch.)

Ex. 417. G is 7·5 miles S. of H, and 10 miles W. of K. What is the distance and bearing of K from H? (Scale 1 mile to 1 cm.)

Ex. 418. X is 17·5 miles N.W. of Y, Y is 23 miles N.E. of Z. What is the distance and bearing of X from Z? (Scale 10 miles to 1 inch.)

Ex. 419.   Two blockhouses are known to be 1000 yards apart, and one of them is due E. of the other.   A party of the enemy are observed by one blockhouse in a N.W. direction, and at the same time by the other in a N.E. direction.   How far are the enemy from each blockhouse?

Ex. 420.   Exeter is 48 miles W. of Dorchester, and Barnstaple is 35 miles N.W. of Exeter.   What is the distance and bearing of Barnstaple from Dorchester?   (Scale 10 miles to 1 in.)

Ex. 421.   From G go 9 miles W. to H, from H go 12 miles N. to A, from A go 17 miles W. to R.   What is the distance from G to R?

Ex. 422.   A is 12 miles N. of H, D is 24 miles S. of H, O is due W. of A, and OH is 42 miles; find OD and OA.

Ex. 423.   XT = 19 miles, MX = 11 miles, MT = 17·5 miles; how far is M from the line XT?

Ex. 424.   Rugby is 44 miles N. of Oxford, and Reading is 24 miles S. 30° E. of Oxford.   Find the distance from Rugby to Reading.   (Scale 10 miles to 1 in.)

Ex. 425.   Southampton is 72 miles S. 53° W. of London, Gloucester is 75° W. of N. from London, and 29° W. of N. from Southampton.   Find the distance between Southampton and Gloucester.   (Scale 10 miles to 1 cm.)

Ex. 426.   P is 64 miles W. of Q, R is due N. of Q; if PR is 72 miles, what is QR?   What is the bearing of P from R?   (Scale 10 miles to 1 cm.)

Ex. 427.   Stafford is 27 miles from Derby and the same distance from Shrewsbury, and the three towns are in a straight line.   Birmingham is 40 miles from Shrewsbury and 35 from Derby.   How far is Stafford from Birmingham?

Ex. 428.   A donkey is tethered to a point 20 feet from a long straight hedge; he can reach a distance of 35 feet from the point to which he is tethered.   How much of the hedge can he nibble?   .

Ex. 429. A and B are two buoys 800 yards apart, B due N. of A. A vessel passes close to B, and steering due E., observes that after 5 minutes the bearing of A is 57° W. of S.   Find the distance the vessel has moved.

Ex. 430. A is a lighthouse.   B and C are two ships 3·5 miles apart. B is due north of A, C due east of B, and C north-east of A.   Find the distance of both ships from the lighthouse.

Ex. 431. Two ships sail from a port, one due N. at 15 miles an hour, the other E.N.E.; at the end of half an hour they are in line with a lighthouse which is 11 miles due E. of the port. At what rate does the second ship sail?

Ex. 432. A man standing on the bank of a river sees a tree on the far bank in a direction 20° W. of N.   He walks 200 yards along the bank and finds that its direction is now N.E.   If the river flows east and west, find its breadth.

Ex. 433. A ferry-boat is moored by a rope 30 yards long to a point in the middle of a river.   The rope is kept taut by the current.   What angle does it turn through as the boat crosses the river, whose width is 30 yards?

Ex. 434. Brixham is 4·6 miles N.E. of Dartmouth, Torquay is 4 miles N. of Brixham, Totnes is 7·4 miles S. 75° W. of Torquay; what is the distance and bearing of Totnes from Dartmouth?

HEIGHTS AND DISTANCES.

If a man who is looking at a tower through a telescope holds the telescope horizontally, and then raises (or "elevates") the end of it till he is looking at the top of the tower, the angle he has turned the telescope through is called the **angle of elevation** of the top of the tower.

If a man standing on the edge of a cliff looks through a horizontal telescope and then lowers (or "depresses") the end of it till he is looking straight at a boat, the angle ho has turned the telescope through is called the **angle of depression** of the boat.

Remember that the angle of elevation and the angle of depression are always angles at the observer's eye.

If O is an observer and A and P two points (see fig. 20), the angle AOP is said to be the **angle subtended** at O by AP.

Ex. 435. In fig. 52, name the angles subtended (i) by BD at A, (ii) by AD at B, (iii) by AC at B.

Ex. 436. A vertical flagstaff 50 feet high stands on a horizontal plane. Find the angles of elevation of the top and middle point of the flagstaff from a point on the horizontal plane 15 feet from the foot of the flagstaff.

Ex. 437. The angle of elevation of the top of the spire of Salisbury Cathedral at a point 1410 feet from its base was found to be 16°. What is the height of the spire?

Ex. 438. A torpedo boat passes at a distance of 100 yards from a fort the guns of which are 100 feet above sea-level; to what angle should the guns be depressed so that they may point straight at the torpedo boat?

Ex. 439. From a point A the top of a church tower is just visible over the roof of a house 50 feet high. If the distance from A to the foot of the tower is known to be 160 yards, and from A to the foot of the house 60 yards, find in feet the height of the tower, and the angle of elevation of its top as seen from A.

Ex. 440. A flagstaff stands on the top of a tower. At a distance of 40 feet from the base of the tower, the angle of elevation of the top of the tower is found to be $23\frac{1}{2}°$, and the flagstaff subtends an angle of $25\frac{1}{2}°$. Find the length of the flagstaff and the height of the tower.

Ex. 441. At two points on opposite sides of a poplar the angles of elevation of its top are 39° and 48°. If the distance between the points is 150 feet, what is the height of the tree?

Ex. 442. From the top of a mast 80 feet high the angle of depression of a buoy is 24°. From the deck it is 5½°. Find the distance of the buoy from the ship.

Ex. 443. At a window 15 feet from the ground a flagstaff subtends an angle of 43°; if the angle of depression of its foot is 11°, find its height.

Ex. 444. A man observes the angle of elevation of the top of a spire to be 23°; he walks 40 yards towards it and then finds the angle to be 29°. What is the height of the spire?

Ex. 445. An observer in a balloon, one mile high, observes the angle of depression of a church to be 35°. After ascending vertically for 20 minutes, he observes the angle of depression to be now 55½°. Find the rate of ascent in miles per hour.

Ex. 446. An observer finds that the line joining two forts A and B subtends a right angle at a point C; from C he walks 100 yards towards B and finds that AB now subtends an angle of 107°; find the distance of A from the two points of observation.

Ex. 447. A man on the top of a hill sees a level road in the valley running straight away from him. He notices two consecutive mile-stones on the road, and finds their angles of depression to be 30° and 13° respectively. Find the height of the hill as a decimal of a mile.

Ex. 448. The shadow of a tree is 30 feet long when the sun's altitude is 59°; find, by drawing, the height of the tree, taking a scale of 1 inch to 10 feet.

Ex. 449. A telegraph pole standing upright on level ground is 23·6 feet high and is partly supported by a wire attached to the top of the pole at one end and fixed to the ground at the other so that its inclination to the pole is 54° 22′.
Find the length of the wire.

Ex. 450. The angle of elevation of the top of a tower on level ground is read off on a theodolite. Find the height of the tower from the following data ·

Reading of theodolite = 15°.

Height of theodolite telescope above ground = 3′ 6″.

Distance of theodolite from foot of tower = 372 yards.

Ex. 451. From a ship at sea the top of Aconcagua has an angle of elevation of 18°. The ship moves out to sea a distance of 5 nautical miles further away from the mountain. The angle of elevation of the top of the mountain is now 13°. Find the height of Aconcagua above sea-level in feet. (1 nautical mile = 6080 ft.)

Ex. 452. Two observations are made to find the height of a certain monument. From the first station the angle of elevation of the top is found to be 32° and from the second station, which is 27 yards from the first and exactly between it and the foot of the monument, the angle of elevation is 43°. If the telescope of the theodolite with which these observations are made is 3 feet above the ground, what is the height of the monument in feet?

Ex. 453. Wishing to find the height of a cliff I fix two marks A and B on the same level in line with the foot of the cliff. From A the angle of elevation of the top of the cliff is 37° and from B the angle of elevation is 23° 30′. If A and B are 120 feet apart, calculate the height of the cliff.

Ex. 454. From a point on a battleship 30 feet above the water, a Torpedo Boat Destroyer is observed steaming away in a straight line. The angle of depression of the bow is observed to be 11°, and that of the stern to be 21°. Find the length of the T. B. D.

Ex. 455. From the top of a mast 70 feet high, two buoys are observed due N. at angles of depression 57° and 37°; find the distance between the buoys to the nearest foot.

Ex. 456. The angles of depression of two boats in a line with the foot of a cliff are 25° 16′ and 38° 39′ as observed by a man at the top of the cliff. If the man is 250 feet above sea-level, find how far apart the boats are.

Ex. 457. A torpedo boat is steering N. 14° E., and from the torpedo boat a lighthouse is observed lying due N. If the speed of the vessel is 15 knots and it passes the lighthouse 40 minutes after the time of observation, find the clearance between the vessel and the lighthouse, and its distance from the lighthouse at the first observation.

Ex. 458. A landmark bears N. 32° W. from a ship. After the ship has sailed 7·2 miles N. 22° E. the landmark is observed to bear N. 71° W. How far is it then from the ship?

Ex. 459. The position of an inaccessible point C is required. From A and B, the ends of a base line 200 yards long, the following bearings are taken:

From A { Bearing of B is N. 70° 30′ E.
          ,,   ,,   C is N. 30° 20′ E.

From B   ,,   ,,   C is N. 59° 40′ W.

Find the distances of C from A and B.

Ex. 460. A ship observes a light bearing N. 52° E. at a distance of 5 miles. She then steams due S. 6 miles, and again observes the light. What does she find the bearing and distance of the light to be at the second observation?

Ex. 461. An Admiral signals to his cruiser squadron (bearing N. 40° W. 50 miles from him) to meet him at a place N. 50° E., 70 miles from his present position. Find bearing and distance of the meeting place from the cruisers.

Ex. 462. A is 1 mile due W. of B.

From A, C bears N. 28° W. and D bears N. 33° E.

From B, C bears N. 34° W. and D bears N. 9° W.

Find the distance and bearing of D from C.

Ex. 463. It is required to find the distance between Stokes Bay Pier and a buoy from the following readings:

Bearing of Stokes Bay Pier from Ryde Pier, N. 9° E.

    ,,   ,, Buoy from Ryde Pier N. 36° W.

    ,,   ,, Buoy from Stokes Bay Pier S. 79° W.

Known distance from Stokes Bay Pier to Ryde Pier, 2·29 m.

Ex. 464. A lies 7 miles N. 32° W. of B; C is 5 miles S. 67° E. of A. Find the distance and bearing of C from B.

Ex. 465. Two rocks A, B are seven miles apart, one being due East of the other. How many miles from each of them is a ship from which it is observed that A bears S. 24° W. and B bears S. 35° E. ?

Ex. 466. From a ship at sea the following observations are made: Dover bears N. 16° E., and Boulogne S. 81° E. From the chart it is found that Dover is 26 miles N. 24° W. of Boulogne. Find the distance of the ship from Dover.

Ex. 467. O and P are points on a straight stretch of shore. P is 4·5 miles N. 74° E. of O. From O a ship at sea bears S. 58° E., and from P the ship bears S. 32° W. Find the distance of the ship from P, and also its distance from the nearest point of the shore.

Ex. 468. A ship steaming due E. at 9·15 knots through the Straits of Gibraltar observes that a point on the Rock bears N. 35° E.; 40 minutes later the same point bears due N.; how far is she from the point at the second observation ?

Ex. 469. A ship is observed to be 3 miles from N. 28° E. a coast-guard station, and to be steaming N. 72° W. After 15 minutes the ship bears N. 36° W. At what rate is she steaming?

Ex. 470. C and D are inaccessible objects. A and B are points 100 yards apart, B due East of A.

From A the bearing of C is due North.

" A " " D is N. 46° E.

" B " " C is N. 63° W.

" B " " D is N. 10° W.

Find (i) distance of C from B,

(ii) distance of D from B,

(iii) distance of C from D.

THIRD STAGE.

BOOK I.

In the Second Stage we arrived at the facts given below. In the Third Stage we shall assume the truth of these facts and deduce other facts or **theorems** from them; we shall not assume the truth of other theorems unless we can prove them theoretically.

Theoretical proof has two advantages over *verification by measurement,* (i) measurement is at best only approximate, (ii) it is impossible to measure every case.

TABLE OF FACTS (OR THEOREMS).

These were enunciated in the Second Stage as the result of processes of intuition and experiment; deductive proofs of them are given in the Appendix, but these are hard and should be postponed.

Angles at a point.

FACT A₁ [Th. I. 1]. If a straight line stands on another straight line, the sum of the two angles so formed is equal to two right angles. (p. 24.)

FACT A₂ [Th. I. 1, Cor.]. If any number of straight lines meet at a point, the sum of all the angles made by consecutive lines is equal to four right angles. (p. 27.)

75

FACT B [Th. I. 2].   If the sum of two adjacent angles is equal to two right angles, the exterior arms of the angles are in the same straight line.   (p. 25.)

FACT C [Th. I. 3].   If two straight lines intersect, the vertically opposite angles are equal.   (p. 28.)

Parallel straight lines.

FACT D [Th. I. 4].   When a straight line cuts two other straight lines, if

  (1) a pair of alternate angles are equal,

  or (2) a pair of corresponding angles are equal,

  or (3) a pair of interior angles on the same side of the cutting line are together equal to two right angles,

then the two straight lines are parallel.   (pp. 30–32.)

FACT E [Th. I. 5].   If a straight line cuts two parallel straight lines,

  (1) alternate angles are equal,

  (2) corresponding angles are equal,

  (3) the interior angles on the same side of the cutting line are together equal to two right angles.   (pp. 30–32.)

Angles of triangle and polygon.

FACT F_1 [Th. I. 8].   The sum of the angles of a triangle is equal to two right angles.   (p. 34.)

FACT F_2 [Th. I. 8, Cor. 1].   If one side of a triangle is produced, the exterior angle so formed is equal to the sum of the two interior opposite angles.   (p. 37.)

FACT G [Th. I. 9].   If the sides of a convex polygon are produced in order, the sum of the angles so formed is equal to four right angles.   (p. 38.)

Congruent triangles.

FACT H [Th. I. 10]. If two triangles have two sides of the one equal to two sides of the other, each to each, and also the angles contained by those sides equal, the triangles are congruent. (p. 49.)

FACT J [Th. I. 11]. If two triangles have two angles of the one equal to two angles of the other, each to each, and also one side of the one equal to the corresponding side of the other, the triangles are congruent. (p. 51.)

FACT K [Th. I. 14]. If two triangles have the three sides of the one equal to the three sides of the other, each to each, the triangles are congruent. (p. 52.)

FACT L [Th. I. 12]. If two sides of a triangle are equal, the angles opposite to these sides are equal. (p. 55.)

FACT M [Th. I. 13]. If two angles of a triangle are equal, the sides opposite to these angles are equal. (p. 56.)

FACT N [Th. I. 15]. If two right-angled triangles have their hypotenuses equal, and one side of the one equal to one side of the other, the triangles are congruent. (p. 59.)

All references to the above will, in future, be to the number of the theorem.

It is so often necessary to prove triangles congruent that it will be well to summarise the facts we use for this purpose.

Two triangles can be proved congruent when we have either of the following sets of equalities given:

(i)   Two sides and the included angle.   (Theorem I. 10.)

(ii)   One side, and any two angles.   (Theorem I. 11.)

(iii)   Three sides.   (Theorem I. 14.)

(iv)   Two sides and a non-included angle provided that angle is a right angle.   (Theorem I. 15.)

RULER AND COMPASS CONSTRUCTIONS.

Hitherto we have constructed our figures with the help of graduated instruments and set square. We shall now make certain constructions with the aid of nothing but a straight edge (not graduated) and a pair of compasses.

We shall use the straight edge

(i) for drawing the straight line passing through any two given points,

(ii) for producing any straight line already drawn.

We shall use the compasses

(i) for describing circles with any given point as centre and radius equal to any given straight line,

(ii) for the transference of distances; i.e. for cutting off from one straight line a part equal to another straight line. [(ii) is really included in (i).]

By means of the facts given on pp. 75–77, we shall show that our constructions are accurate.

In the exercises, when you are asked to construct a figure, you should always explain your construction in words. You need not give a proof unless you are directed to do so.

In the earlier constructions the figures are shown with

given lines—thick,

construction lines—fine,

required lines—of medium thickness,

lines needed only for the proof—broken.

In making constructions, only the necessary parts of construction circles should be drawn even though "the circle" is spoken of.

Through a point O in a straight line OX to draw a straight line OY so that ∠XOY may be equal to a given angle BAC.

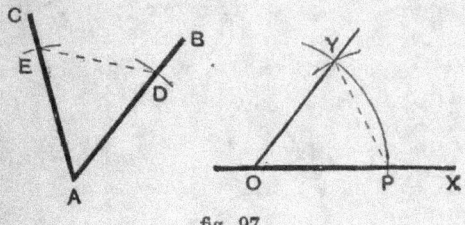

fig. 97.

Construction With centre A and any radius describe a circle cutting AB, AC at D, E respectively.

With centre O and the same radius describe a circle PY cutting OX at P.

With centre P and radius = DE describe a circle cutting the circle PY at Y. Join OY.

Then ∠XOY = ∠BAC.

Proof Join DE and PY.

In the △ˢ OPY, ADE,

$$OP = AD, \quad OY = AE, \quad PY = DE, \qquad Constr.$$
$$\therefore \triangle POY \equiv \triangle ADE, \qquad \text{I. 14.}$$
$$\therefore \angle POY = \angle DAE,$$
$$\text{i.e. } \angle XOY = \angle BAC.$$

[*The protractor must not be used in Exx.* 471–474.]

Ex. **471.** Draw an acute angle and construct an equal angle.

Ex. **472.** Draw an obtuse angle and make a copy of it.

Ex. **473.** Draw a triangle ABC; at a point O make a copy of its angles in the manner of fig. 51.

Ex. **474.** Draw a straight line EF and mark a point G (about 2 in. from the line); through G draw a line parallel to EF.

[Draw any line through G cutting EF at H; make ∠HGC=∠GHF; see fig. 48.]

To bisect a given angle.

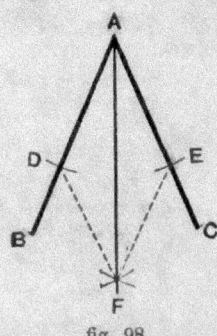

fig. 98.

Let BAC be the given angle.

Construction From AB, AC cut off equal lengths AD, AE.

With centres D and E and any convenient radius describe equal circles intersecting at F.

Join AF.

Then AF bisects ∠ BAC.

Proof Join DF and EF.

In the △ˢ ADF, AEF,

$$\left\{ \begin{array}{l} AD = AE, \\ DF = EF, \\ AF \text{ is common.} \end{array} \right. \qquad \begin{array}{l} Constr. \\ \text{,,} \\ \ \end{array}$$

∴ △ ADF ≡ △ AEF,   I. 14.

∴ AF bisects ∠ BAC.

"*Any convenient radius.*" If it is found that the equal circles do not intersect, the radius chosen is not convenient, for the construction breaks down; it is necessary to take a larger radius so that the circles may intersect.

For exercises on this construction, see p. 82.

To draw the perpendicular bisector of a given straight line.

To bisect a given straight line.

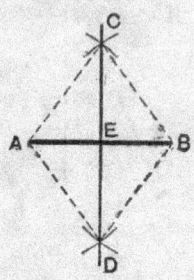

fig. 99.

Let AB be the given straight line.

Construction   With centres A and B and any convenient radius describe equal circles intersecting at C and D.

Join CD and let it cut AB at E.

Then CD is the perpendicular bisector of AB, and E is the mid-point of AB.

Proof   Join AC, AD, BC, BD.

In the △ˢ ACD, BCD,

$\begin{cases} AC = BC, \\ AD = BD, \\ CD \text{ is common,} \end{cases}$ *Constr.*
                                                                                                    ,,

∴ △ ACD ≡ △ BCD,   I. 14.

∴ ∠ ACD = ∠ BCD.

In the △ˢ ACE, BCE,

$\begin{cases} AC = BC, \\ CE \text{ is common,} \\ \angle ACE = \angle BCE, \end{cases}$ *Constr.*

                                                                                                    Proved

∴ △ ACE ≡ △ BCE,   I. 10.

∴ AE = BE,

and ∠ˢ CEA, CEB are equal and are therefore rt. ∠ˢ,   *Def.*

∴ CD bisects AB at right angles.

¶Ex. 475. If fig. 98 were folded about AF, what points would coincide? What lines?

¶Ex. 476. Make two equal angles and bisect them; in one case join the vertex to the nearer point at which the equal circles intersect, in the other to the further point.

Which gives the better result?

¶Ex. 477. Is there any case in which one point of intersection would coincide with the vertex of the angle?

Ex. 478. Make an angle of (i) 60°, (ii) 120° (without protractor or set square).

Ex. 479. Construct angles of 15°, 30° and 150° (without protractor or set square).

Ex. 480. Draw a large triangle and bisect each of its angles.

Ex. 481. Draw a straight line and bisect it.

Ex. 482. Draw a large acute-angled triangle; draw the perpendicular bisectors of its three sides.

Ex. 483. Repeat Ex. 482 for (i) a right-angled triangle, (ii) an obtuse-angled triangle.

Ex. 484. Draw any chord of a circle and its perpendicular bisector.

DEF. The straight line joining a vertex of a triangle to the mid-point of the opposite side is called a median.

Ex. 485. Draw a large triangle; and draw its three medians. Are the angles bisected?

¶Ex. 486. Call one of the short edges of your paper AB ; construct its perpendicular bisector by folding. Fold the paper again so that the new crease may pass through A, and B may fall on the old crease; mark the point C on which B falls and join CA, CB. What kind of triangle is ABC?

Ex. 487. Describe a circle and on it take three points A, B, C; join BC, CA, AB. Bisect angle BAC and draw the perpendicular bisector of BC. Produce the two bisectors to meet.

¶ Ex. 488. Draw a large obtuse angle (very nearly 180°) and bisect it.

To draw a straight line perpendicular to a given straight line AB from a given point P in AB.

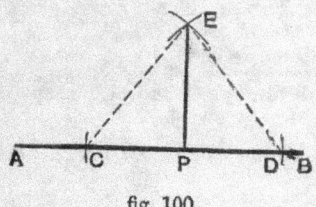

fig. 100.

Construction   From PA, PB cut off equal lengths PC, PD.

With centres C and D and any convenient radius describe equal circles intersecting at E.

Join PE.

Then PE is ⊥ to AB.

Proof   Join CE, DE.

In the △ˢ CPE, DPE,

$$\begin{cases} \text{PC} = \text{PD}, & \textit{Constr.} \\ \text{CE} = \text{DE (radii of equal} \odot^s\text{)}, \\ \text{PE is common.} \end{cases}$$

∴ △ CPE ≡ △ DPE,   I. 14.

∴ ∠ EPC = ∠ EPD,

∴ PE is ⊥ to AB.   *Def.*

[*The protractor and set square must not be used in the constructions of Exx.* 489, 490.]

Ex. 489. Draw a straight line, and a straight line at right angles to it. Test with set square.

Ex. 490. Construct angles of 45° and 75°.

To draw a straight line perpendicular to a given straight line AB from a given point P outside AB.

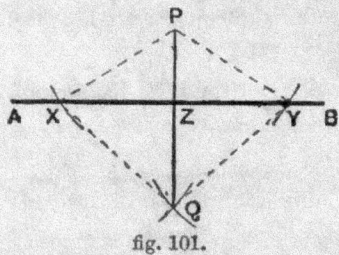

fig. 101.

Construction   With centre P and any convenient radius describe a circle cutting AB at X and Y.

With centres X and Y and any convenient radius describe equal circles intersecting at Q.   Join PQ cutting AB at Z.

Then PZ is ⊥ to AB.

Proof   Join XP, XQ, YP, YQ.

In the △ˢ PQX, PQY,

$\begin{cases} PX = PY \text{ (radii of a } \odot), \\ QX = QY \text{ (radii of equal } \odot^s), \\ PQ \text{ is common.} \end{cases}$

∴ △ PQX ≡ △ PQY,   I. 14.

∴ ∠ XPQ = ∠ YPQ.

We can now prove that

△ PXZ ≡ △ PYZ, (give the three reasons)

∴ ∠ PZX = ∠ PZY,

∴ PZ is ⊥ to AB.

Ex. **491.** Draw a large acute-angled triangle ; from each vertex draw a perpendicular to the opposite side.

Ex. **492.** Repeat Ex. 491 with a right-angled triangle.

Ex. **493.** Repeat Ex. 491 with an obtuse-angled triangle. [You will have to produce two of the sides.]

Ex. **494.** From the centre of a circle drop a perpendicular on a chord of the circle.

CONSTRUCTION OF TRIANGLES FROM GIVEN DATA.

We have seen how to construct triangles having given

(i)   the three sides (p. 45);

(ii) two sides and the included angle (p. 43);

(iii) one side and two angles (p. 44).

In Ex. 259, we saw that, given the angles, it is possible to construct an unlimited number of different triangles.

If two angles of a triangle are given, the third angle is known; hence the three angles do not constitute more than two data.

We have still to consider the case in which two sides are given and an angle not included by these sides.

¶Ex. 495. Construct a triangle ABC having given BC = 2·4 in., CA = 1·8 in., and ∠ B = 32°.

First make BC = 2·4 in. and ∠ CBD = 32°.

A must lie somewhere on BD, and must be 1·8 in. from C.

Where do all the points lie which are 1·8 in. from C?

How many points are there which are on BD and also 1·8 in. from C?

We see that it is possible to construct two unequal triangles which satisfy the given conditions. This case is therefore called the **ambiguous case**.

Ex. **496.** Construct triangles to the following data:—

(i)   BC = 8·7 cm.,   CA = 5·3 cm.,   ∠ B = 29°;

(ii) BC = 7·3 cm.,   CA = 9·0 cm.,   ∠ A = 53°;

(iii) AB = 3·9 in.,   AC = 2·6 in.,   ∠ C = 68°;

(iv) AB = 2·2 in.,   BC = 3·7 in.,   ∠ A = 90°;

(v)   AC = 5·3 cm.,   BC = 10 cm.,   ∠ B = 32°;

(vi) AC = 1·6 in.,   BC = 4·7 in.,   ∠ B = 26°.

†Ex. **497.** Prove (theoretically) that the two triangles obtained in Ex. 496 (iv) are congruent.

Ex. **498.** Describe a right-angled triangle having given its hypotenuse and one acute angle.

Ex. **499.** Show how to construct a triangle ABC having AB=3 in., BC=5 in., and the median to BC=2·5 in.

Ex. **500.** Show how to construct a triangle ABC having given AB=10 cm., AC=8 cm., and the perpendicular from A to BC=7·5 cm. Is there any ambiguity?

[First draw the line of the base, and the perpendicular.]

Ex. **501.** Show how to construct a triangle ABC having given AB=11·5 cm., BC=4·5 cm., and the perpendicular from A to BC=8·5 cm. Is there any ambiguity?

¶ Ex. **502.** Show how to construct a quadrilateral having given its sides and one of its angles.   Is there any ambiguity?

CONTINUOUS CHANGE OF A FIGURE.

The student should accustom himself to thinking of figures changing their dimensions continuously; he should vary some dimensions and trace how other dimensions change in consequence; in some cases a graph might be drawn.

For instance, let O be a fixed point outside a fixed straight line; let P be a point moving continuously along the straight line; trace the change in the length of OP.

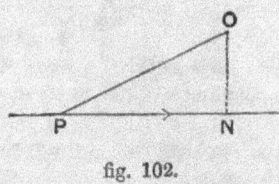

fig. 102.

If P lies a long way to the left, OP is great; as P moves to the right, OP becomes smaller and smaller, until it reaches its least, or **minimum**, length when OP coincides with the perpendicular ON. When P passes N, OP begins to increase again; and as P moves further and further to the right, OP increases without limit.

Since the perpendicular is the shortest line that can be drawn from a given point to a given straight line, it is called the **distance** of the point from the line. Remember that the distance of a point from a straight line is always to be measured perpendicular to the line.

¶Ex. **503.** X'OX is a fixed line of unlimited length; a line OP of fixed length rotates about O; PN is drawn ⊥ to X'OX. Trace the change in PN as ∠XOP increases from 0° to 180°. Show the changes in PN by means of a graph. (Represent degrees horizontally; mark values of PN for 0°, 10°, 20°, etc.)

¶Ex. **504.** The sides AB, AC of △ABC are of constant lengths; trace the change in BC as ∠BAC increases from 0° to 180°.

¶Ex. **505.** The base and height of a triangle are of constant lengths; how may the vertex move?
Trace the consequent change in (i) the vertical angle, (ii) the remaining sides.

¶Ex. **506.** AB and ∠BAC (acute) are fixed dimensions of a △ABC, and the side AC increases continuously from zero; trace the change in the length of BC.

PARALLELOGRAMS.

¶Ex. **507.** Make an angle ABC=65°, cut off BA=2·2 in., BC=1·8 in.; through A draw AD parallel to BC, through C draw CD parallel to BA.

DEF. A quadrilateral with its opposite sides parallel is called a **parallelogram**.

¶Ex. **508.** What relations do you notice between any of the sides and angles of the parallelogram you constructed in Ex. 507?

THEOREM 22.

(1) The opposite angles of a parallelogram are equal.

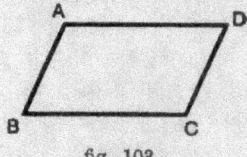

fig. 103.

Data   ABCD is a parallelogram.

To prove that $\angle A = \angle C$, $\angle B = \angle D$.

Proof   Since AD and BC are ∥, and AB meets them,

$\therefore\ \angle A + \angle B = 2$ rt. $\angle$ s.   I. 5.

Sim^{ly} $\angle B + \angle C = 2$ rt. $\angle$ s.   I. 5.

$\therefore\ \angle A + \angle B = \angle B + \angle C$,

$\therefore\ \angle A = \angle C$.

Sim^{ly} $\angle B = \angle D$.   Q. E. D.

(2) The opposite sides of a parallelogram are equal.

(3) Each diagonal bisects the parallelogram.

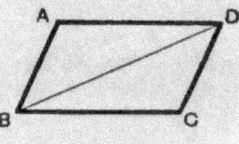

fig. 104.

Data   ABCD is a parallelogram, and BD one of its diagonals.

To prove that   AB = CD, AD = CB, and that BD bisects the parallelogram.

Proof   [△s ABD, CDB must be proved congruent.]

Since AD is ∥ to BC and BD meets them,

$\angle$ ADB = alt. $\angle$ CBD.   I. 5.

Since AB is ∥ to CD and BD meets them,

$\angle$ ABD = alt. $\angle$ CDB.   I. 5.

∴ in △s ABD, CDB,

{ ∠ADB = ∠CBD,
  ∠ABD = ∠CDB,
  BD is common,

∴ △ABD ≡ △CDB,   I. 11.

∴ AB = CD,   AD = CB.

And since △ABD ≡ △CDB,

BD bisects the parallelogram.

Sim^ly AC bisects the parallelogram.   Q. E. D.

(4) **The diagonals of a parallelogram bisect one another.**

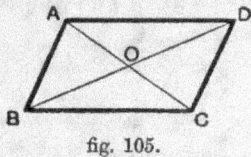

fig. 105.

Data ABCD is a parallelogram; its diagonals AC, BD intersect at O.

To prove that   OA = OC and OD = OB.

Proof   Since AD is ∥ to BC and BD cuts them,

∴ ∠ADO = ∠CBO,

∴ in △s OAD, OCB

{ ∠ADO = ∠CBO,
  ∠AOD = vert. opp. ∠COB,
  AD = CB,   I. 22 (2).

∴ the △s are congruent,   I. 11.

∴ OA = OC and OD = OB.   Q. E. D.

COR. 1. **Parallel straight lines are everywhere equidistant*.**

COR. 2. **If a parallelogram has one of its angles a right angle, all its angles must be right angles.**

COR. 3. **If one pair of adjacent sides of a parallelogram are equal, all its sides are equal.**

* For distance of a point from a line, see p. 87.

†Ex. **509.**   Prove Cor. 1.

†Ex. **510.**   Prove Cor. 2.

†Ex. **511.**   Prove Cor. 3.

DEF.   A parallelogram which has one of its angles a right angle is called a **rectangle.**

Cor. 2 proves that all the angles of a rectangle are right angles.

DEF.   A rectangle which has two adjacent sides equal is called a **square.**

Cor. 3 proves that all the sides of a square are equal to one another. Again, since a square is a rectangle, all its angles are right angles.

DEF.   A parallelogram which has two adjacent sides equal is called a **rhombus.**

Cor. 3 proves that all the sides of a rhombus are equal to one another.

DEF.   A quadrilateral which has only one pair of sides parallel is called a **trapezium.**

DEF.   A trapezium in which the sides which are not parallel are equal to one another is called an **isosceles trapezium.**

¶Ex. **512.**   Make a parallelogram with strips of wood or cardboard hinged at the corners; use elastic for the diagonals.

As the figure changes shape, trace the change in the lengths of the diagonals. Are they ever equal? What are their maximum and minimum lengths?

Trace the change in the angle between the diagonals. Can it ever be a right angle?

Trace the change in the height. What are its maximum and minimum values?

How would the above be modified if the parallelogram were a rhombus?

¶Ex. **513.**   The base and height of a parallelogram remain constant; trace the change in (i) the other side, (ii) the diagonals.

¶Ex. **514.**   The base of a rectangle remains constant; trace the change in the angle between the diagonals as the height increases from zero.

¶Ex. 515. Draw a parallelogram ABCD; from AB, AD cut off equal lengths AX, AY; through X, Y draw parallels to the sides to meet the opposite sides. Indicate what lines and angles are equal.

Ex. 516. Copy the table given below; indicate for which figures the given properties are always true by inserting the words "yes" or "no" in the corresponding spaces.

	Opposite sides and angles equal	Diagonals bisect one another	Angles at corners right angles	Diagonals equal	Diagonals at right angles	Adjacent sides equal	Diagonals bisect angles
Parallelogram							
Rectangle							
Square							
Rhombus							

†Ex. 517. In fig. 123, ABCD is a parallelogram and PBCQ is a rectangle; prove that △ BPA ≡ △ CQD.

†Ex. 518. The bisectors of two adjacent angles of a parallelogram are at right angles to one another.

†Ex. 519. The bisectors of two opposite angles of a parallelogram are parallel.

†Ex. 520. Any straight line drawn through O, in fig. 105, and terminated by the sides of the parallelogram is bisected at O.

†Ex. 521. Draw an isosceles triangle ABC and a line parallel to the base cutting the sides in D, E; prove that DECB is an isosceles trapezium.

†Ex. **522.**   ABCD is an isosceles trapezium (AD = BC); prove that

$$\angle C = \angle D.$$

[Through B draw a parallel to AD.]

†Ex. **523.**   If in Ex. 522 E, F are the mid-points of AB, CD, then EF is perpendicular to AB.   [Join AF, BF.]

NOTE ON A THEOREM AND ITS CONVERSE.

The enunciation of a theorem can generally be divided into two parts (i) the data, or hypothesis, (ii) the conclusion.

For instance in Theorem 22 (1) the data is that the figure is a parallelogram; the conclusion is that the opposite angles are equal.

If the data and conclusion are interchanged a second theorem is obtained, which is called the **converse** of the first theorem.

Thus the converse of I. 22 (1) is as follows:—if the opposite angles of a quadrilateral are equal, it is a parallelogram.   This is Theorem I. 23 (1).

It must not be assumed that the converses of all true theorems are true; e.g. "if two angles are vertically opposite, they are equal" is a true theorem, but the converse "if two angles are equal, they are vertically opposite" is not a true theorem.

¶Ex. **524.**   State the converses of the following : are they true?

(i)   If two sides of a triangle are equal, then two angles of the triangle are equal.

(ii)   If a triangle has one of its angles a right angle, two of its angles are acute.

(iii)   London Bridge is a stone bridge.

(iv)   A nigger is a man with woolly hair.

Theorem 23.†

[Converses of Theorem 22.]

(1) A quadrilateral is a parallelogram if both pairs of opposite angles are equal.

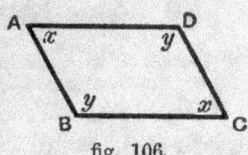

fig. 106.

Data   ABCD is a quadrilateral in which

$$\angle A = \angle C = \angle x \text{ (say) and } \angle B = \angle D = \angle y \text{ (say).}$$

To prove that   ABCD is a parallelogram.

Proof The sum of the angles of a quadrilateral is equal to
      4 rt. ∠s,   I. 9, *Cor.*

$$\therefore 2\angle x + 2\angle y = 4 \text{ rt. } \angle s,$$

$$\therefore \angle x + \angle y = 2 \text{ rt. } \angle s,$$

$$\therefore \angle A + \angle B = 2 \text{ rt. } \angle s,$$

$$\therefore \text{ AD is } \| \text{ to BC.} \qquad \text{I. 4.}$$

Also $\angle A + \angle D = 2$ rt. ∠s,

$$\therefore \text{ AB is } \| \text{ to DC,} \qquad \text{I. 4.}$$

$$\therefore \text{ ABCD is a } \|^{\text{ogram}}. \qquad \text{Q. E. D.}$$

(2) A quadrilateral is a parallelogram if one pair of opposite sides are equal and parallel.

(Draw a diagonal and prove the two triangles congruent.)

(3) A quadrilateral is a parallelogram if both pairs of opposite sides are equal.

(Draw a diagonal and prove the two triangles congruent.)

(4)   **A quadrilateral is a parallelogram if its diagonals bisect one another.**

(Prove two opposite triangles congruent.)

Cor.   **If equal perpendiculars are erected on the same side of a straight line, the straight line joining their extremities is parallel to the given line.**

†Ex. **525.**   Prove I. 23 (2).

†Ex. **526.**   Prove I. 23 (3).

†Ex. **527.**   Prove I. 23 (4).

†Ex. **528.**   Prove the Corollary.

†Ex. **529.**   Show how to construct, without using set square, a parallelogram having given two adjacent sides and the angle between them. Give a proof.

Ex. **530.**   Show how to construct a square on a given straight line.

Ex. **531.**   Show how to construct on a given straight line a rhombus having one of its angles = 60° (without protractor or set square).

†Ex. **532.**   Show how to construct a parallelogram having given two sides and a diagonal. Give a proof.

Ex. **533.**   Show how to construct a rectangle given one side and a diagonal.

Ex. **534.**   Show how to construct a parallelogram given the diagonals and the angle between them.

Ex. **535.**   Show how to construct a rhombus given the lengths of the diagonals.

Ex. **536.**   Show how to construct an isosceles trapezium whose sides are 4 in., 3 in., 1·5 in., 1·5 in.

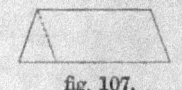

fig. 107.

†Ex. **537.**   The straight line joining the mid-points of two opposite sides of a parallelogram is parallel to the other two sides.

†Ex. **538.** ABCD is a parallelogram; AB, CD are bisected at X, Y respectively; prove that BXDY is a parallelogram.

†Ex. **539.** If the diagonals of a quadrilateral are equal and bisect one another at right angles, the quadrilateral must be a square.

†Ex. **540.** Two straight lines bisect one another at right angles; prove that they are the diagonals of a rhombus.

†Ex. **541.** If the diagonals of a parallelogram are equal, it must be a rectangle.

†Ex. **542.** An equilateral four-sided figure with one of its angles a right angle must be a square.

†Ex. **543.** In a quadrilateral ABCD, $\angle A = \angle B$ and $\angle C = \angle D$; prove that ABCD is an isosceles trapezium. In what case would it be a parallelogram?

A practical method of drawing a straight line parallel to a given straight line and at a given distance from it is as follows:— take two points on the line (as far apart as possible), with these points as centres describe circles with the given distance as radius, then with the ruler draw a common tangent to the circles.

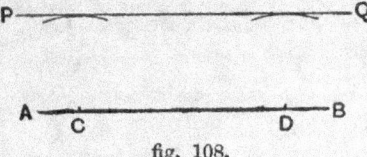

fig. 108.

Ex. **544.** On a base 3 in. long construct a parallelogram of height 1·2 in. with an angle of 55°. Measure the other side.

Ex. **545.** Construct a rhombus whose side is 7·3 cm., the distance between a pair of opposite sides being 5·6 cm. Measure its acute angle.

¶Ex. **546.** Draw a straight line and cut off from it two equal parts AC, CE; through A, C, E draw three parallel straight lines and draw a line cutting them at B, D, F; measure BD, DF. (See fig. 109.)

Theorem 24.

If there are three or more parallel straight lines, and the intercepts made by them on any one straight line that cuts them are equal, then the corresponding intercepts on any other straight line that cuts them are also equal.

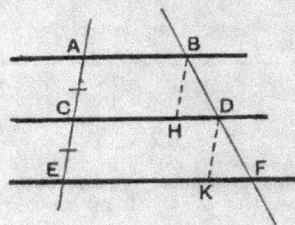

fig. 109.

Data   The parallels AB, CD, EF are cut by the straight lines ACE, BDF, and the intercepts AC, CE are equal.

To prove that   the corresponding intercepts BD, DF are equal.

Construction   Through B draw BH ‖ to ACE to meet CD at H.

               Through D draw DK ‖ to ACE to meet EF at K.

Proof   [△s BHD, DKF must be proved congruent.]

               AH is a ‖ᵒᵍʳᵃᵐ, ∴ AC = BH,   I. 22.

               CK is a ‖ᵒᵍʳᵃᵐ, ∴ CE = DK.

                  But AC = CE,   *Data*

                  ∴ BH = DK.

               Now CD is ‖ to EF,

            ∴ ∠ BDH = corresp. ∠ DFK.   I. 5.

         Again BH, DK are ‖ (each ‖ to ACE),

            ∴ ∠ DBH = corresp. ∠ FDK,   I. 5.

          ∴ in △s BHD, DKF

              { ∠ BDH = ∠ DFK,

               ∠ DBH = ∠ FDK,

                  BH = DK,

          ∴ the △s are congruent,

              ∴ BD = DF.   I. 11.

                               Q. E. D.

To divide a given straight line into five equal parts.

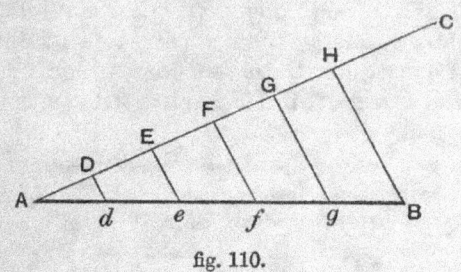

fig. 110.

Let AB be the given straight line.

Construction   Through A draw AC making any angle with AB.

From AC cut off any part AD.

From DC cut off parts DE, EF, FG, GH, equal to AD, so that AH is five times AD.

Join BH.

Through D, E, F, G draw st. lines ∥ to BH.

Then AB is divided into 5 equal parts.

Proof   AD = DE = ...,   *Constr.*

and Dd, Ee, ..., HB are all parallel.   *Constr.*

∴ Ad = de = ...,   I. 24.

∴ AB is divided into 5 equal parts.

Ex. **547.**   Trisect a given straight line by eye; check by making the construction.

Ex. **548.**   Divide a straight line of 10 cm. into six equal parts; check.

Ex. **549.**   **From a given straight line cut off a part equal to $\frac{2}{5}$ of the whole line.**

Ex. **550.**   **Divide a straight line of 13·3 cm. in the ratio of 3 : 4.**

[Divide the straight line (AB) into seven (i.e. 3 + 4) equal parts; if D is the third point of division from A, AD contains three parts and DB contains four parts, ∴ $\frac{AD}{DB} = \frac{3}{4}$.]

¶Ex. **551.** Show how to divide a given straight line into any required number of equal parts (say 6) by fitting over it a piece of tracing paper ruled with equidistant lines.

---

†Ex. **552.** The straight line drawn through the mid-point of one side of a triangle parallel to the base bisects the other side.

[Let A, B coincide in fig. 109.]

†Ex. **553. The straight line joining the mid-points of the sides of a triangle is parallel to the base.**

[Prove this with the following construction:—

Let ABC be the triangle; D, E the mid-points of AB, AC. Produce DE to F, and draw CF parallel to BA.]

†Ex. **554. The straight line joining the mid-points of the sides of a triangle is equal to half the base.**

[Join the mid-point of the base to the mid-point of one of the sides.]

†Ex. **555.** The straight lines joining the mid-points of the sides of a triangle divide it into four congruent triangles.

†Ex. **556.** Show how to construct a triangle, given the three mid-points of its sides. Give a proof.

†Ex. **557.** If AD = $\frac{1}{4}$AB and AE = $\frac{1}{4}$AC, prove that DE is parallel to BC and equal to a quarter of BC.

†Ex. **558.** If the mid-points of the adjacent sides of a quadrilateral are joined, the figure thus formed is a parallelogram.

[Draw a diagonal of the quadrilateral.]

†Ex. **559.** The straight lines joining the mid-points of opposite sides of a quadrilateral bisect one another.

LOCI.

Mark two points A and B, 2 inches apart. Mark a point 3 inches from A and also 3 inches from B: then a point 4 inches from A and B.

In a similar way mark about 10 points equidistant from A and B; some above and some below AB.

Notice what pattern this set of points seems to form. Draw a line passing through all of them.

Find a point on AB equidistant from A and B; this belongs to the set of points.

The pattern formed by all possible points equidistant from two fixed points A and B is called the **locus** of points equidistant from A and B.

¶Ex. **560.** What is the locus of points at a distance of 1 inch from a fixed point O?

¶Ex. **561.** Draw a straight line right across your paper. Construct the locus of points distant 1 inch from this line.

(Do this either by marking a number of such points; or, if you can, without actually marking the points. Remember that the distance is reckoned perpendicular to the line.)

¶Ex. **562.** A bicyclist is riding straight along a level road. What is the locus of the hub of the back wheel?

¶Ex. **563.** What is the locus of the tip of the hand of a clock?

¶Ex. **564.** What is the locus of a man's hand as he works the handle of a common pump?

¶Ex. **565.** A stone is thrown into still water and causes a ripple to spread outwards. What is the locus of the points which the ripple reaches after one second?

¶Ex. **566.** Sound travels about 1100 feet in a second. A gun is fired; what is the locus of all the people who hear the sound 1 second later?

¶Ex. **567.** A round ruler rolls down a sloping plank; what is the locus of the centre of one of the ends of the ruler?

¶Ex. **568.** A man walks along a straight road, so that he is always equidistant from the two sides of the road. What is his locus?

¶Ex. **569.** A runner runs round a circular racing-track, always keeping one yard from the inner edge. What is his locus?

¶Ex. **570.** Two coins are placed on a table with their edges in contact. One of them is held firm, and the other rolls round the circumference of the fixed coin. What is the locus of the centre of the moving coin? Would the locus be the same if there were slipping at the point of contact?

¶Ex. **571.** What is the locus of a door-handle as the door opens?

¶Ex. **572.** What is the locus of a clock-weight as the clock runs down?

¶Ex. 573. Slide your set-square round on your paper, so that the right angle always remains at a fixed point. What are the loci of the other two vertices?

The above exercises suggest the following definition of a locus.

DEF.   If a point moves so as to satisfy certain conditions the **path** traced out by the point is called its **locus**.

Ex. 574. A man stands on the middle rung of a ladder against a wall. The ladder slips down; find the locus of the man's feet.

(Do this by drawing two straight lines at right angles to represent the wall and the ground; take a length of, say, 4 inches to represent the ladder; draw a considerable number of different positions of the ladder as it slips down; and mark the middle points. This is called **plotting** a locus.

The exercise is done more easily by drawing the ladder (the line of 4 inches) on transparent **tracing-paper**; then bring the ends of the ladder on to the two lines of the paper below; and prick through the middle point.)

¶Ex. 575. Draw two unlimited lines, intersecting near the middle of your paper at an angle of 60°. By eye, mark a point equidistant from the two lines. Mark a number of such points, say 20, in various positions. The pattern formed should be *two* straight lines. How are these lines related to the original lines? How are they related to one another?

¶Ex. 576. (On squared paper.) Draw a pair of lines at right angles (OX, OY); plot a series of points each of which is twice as far from OX as from OY. What is the locus? (Keep your figure for the next Ex.)

Ex. 577. Using the figure of Ex. 576, plot the locus of points 3 times as far from OX as from OY; also the locus of points $\frac{1}{2}$ as far from OX as from OY.

Ex. 578. (On squared paper.) Plot the locus of a point which moves so that the sum of its distances from two lines at right angles is always 4 inches.

Ex. 579. (On squared paper.) Plot the locus of a point which moves so that the difference of its distances from two lines at right angles is always 1 inch.

Ex. 580. Draw a line, and mark a point O 2 inches distant from the line. Let P be a point moving along the line. Experimentally, plot the locus of the mid-point of OP.

Ex. 581. A point O is 3 cm. from the centre of a circle of radius 5 cm. Plot the locus of the mid-point of OP, when P moves round the circumference of the circle.

THEOREM 25.

The locus of a point which is equidistant from two fixed points is the perpendicular bisector of the straight line joining the two fixed points.

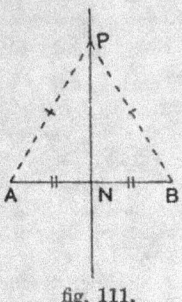

fig. 111.

Data   P is any one position of a point which is always equidistant from two fixed points A and B.

To prove that   P lies on the perpendicular bisector of AB.

Construction   Join AB; let N be the middle point of AB.

Join NP.

Proof   In the △s ANP, BNP,

$\begin{cases} AP = BP, \\ AN = BN, \\ PN \text{ is common,} \end{cases}$ *Data*
                                        Constr.

∴ the triangles are congruent,   I. 14.

∴ ∠ ANP = ∠ BNP,

∴ PN is ⊥ to AB,

∴ P lies on the perpendicular bisector of AB.

Sim^{ly} it may be shown that any other point equidistant from A and B lies on the perpendicular bisector of AB.

Q. E. D.

NOTE.   It will be noticed that N is a point on the locus.

†Ex. **582.**   Prove that any point on the perpendicular bisector of a line AB is equidistant from A, B.

THEOREM 26.

The locus of a point which is equidistant from two intersecting straight lines consists of the pair of straight lines which bisect the angles between the two given lines.

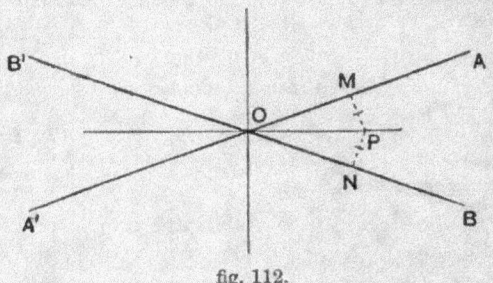

fig. 112.

Data   AOA', BOB' are two intersecting straight lines; P is any one position (in ∠ AOB, say) of a point which is always equidistant from AOA', BOB'.

To prove that   P lies on one of the bisectors of the angles formed by AOA', BOB'.

Construction   Draw PM, PN ⊥ to AA', BB' respectively.
Join OP.

Proof   In the rt. ∠d △s POM, PON,

       ⎰ ∠s M and N are rt. ∠s,   *Constr.*

       ⎨   OP is common,

       ⎱   PM = PN,   *Data*

     ∴ the triangles are congruent,   I. 15.

          ∴ ∠ POM = ∠ PON,

     ∴ P lies on the bisector of ∠ AOB (or ∠ A'OB').

Sim^ly, if P be taken in ∠ AOB' or ∠ A'OB, it may be shown that the point lies on the bisector of ∠ AOB' (or ∠ A'OB).

Q. E. D.

†Ex. **583**. Prove that any point on the bisector of an angle is equidistant from the arms of that angle.

†Ex. **584**. Prove formally that the locus of points at a distance of 1 inch from a given line, on one side of it, is a parallel line. (Take two such points, and show that the line joining them is parallel to the given line.)

†Ex. **585**. O is a fixed point. P moves along a fixed-line; Q is in OP produced, and PQ=OP. Prove that the locus of Q is a parallel line.

INTERSECTION OF LOCI.

Draw two unlimited straight lines AOA′, BOB′, intersecting at an angle of 45°. It is required to find a point (or points) distant 1 inch from each line.

First draw the locus of points distant 1 inch from AOA′; this consists of a pair of lines parallel to AOA′ and distant 1 inch from it. The points we are in search of must certainly lie somewhere upon this locus.

Next draw the locus of points distant 1 inch from BOB′. The required points must lie upon this locus also.

The two loci will be found to intersect in four points. These are the points required.

Ex. **586**. Draw two unlimited straight lines intersecting at an angle of 80°. Find a point (or points) distant 2 cm. from the one line and 4 cm. from the other.

Ex. **587**. Draw an unlimited straight line and mark a point O 2 inches from the line. Find a point (or points) 3 inches from O and 3 inches from the line. (What is the locus of points 3 inches from O? What is the locus of points 3 inches from the line? Draw these loci.) Measure the distance between the two points found.

Ex. **588**. In Ex. 587 find two points distant 4 inches from O and from the line. Measure the distance between them.

Ex. **589**. In Ex. 587 find as many points as you can distant 1 inch from both point and line.

Ex. **590**. If A, B are two points 3 inches apart, show how to find a point (or points) distant 4 inches from A and 5 inches from B.

Ex. **591.**   Make an angle of 45°; on one of the arms mark a point A 3 inches from the vertex of the angle.   Find a point (or points) equidistant from the arms of the angle, and 2 inches from A.   Measure the distance between the two points found.

Ex. **592.**   On the diagonal BD (produced if necessary) of a quadrilateral ABCD, show how to find a point

    (1)   equidistant from A and C,
    (2)   equidistant from AB and AD,
    (3)   equidistant from AB and DC.

Ex. **593.**   Find a point on the base of an equilateral triangle (side 10 cm.) which is 4 cm. from one side.   Measure the two parts into which it divides the base.

Ex. **594.**   Show how to find a point on the base of a scalene triangle equidistant from the two sides.   Is this the middle point of the base?

Ex. **595.**   Draw a circle, a diameter AB, and a chord AC through A. Show how to find a point P on the circle equidistant from AB and AC.

Ex. **596.**   Show how to find a point (or points) on a given circle equidistant from a pair of parallel straight lines.

Ex. **597.**   Show how to find a point (or points) equidistant from two given straight lines and at a given distance from a third line.

Ex. **598.**   Show how to find a point (or points) equidistant from two sides of a triangle and also equidistant from the ends of the third side.

Ex. **599.**   Draw a triangle ABC; find a point O which is equidistant from B, C; and also equidistant from C, A.   Test by drawing circle with centre O to pass through A, B, C.

Ex. **600.**   Draw a triangle ABC.   Inside the triangle find a point P which is equidistant from AB and BC; and also equidistant from BC and CA. From P draw perpendiculars to the three sides; with P as centre and one of the perpendiculars as radius draw a circle.

Ex. **601.**   A river with straight banks is crossed, slantwise, by a straight weir.   Sketch a figure representing the position of a boat which finds itself at the same distance from the weir and the two banks.

†Ex. **602.**   P is a moving point on a fixed line AB; O is a fixed point outside the line.   P is joined to O, and PO is produced to Q so that OQ=PO.   Prove that the locus of Q is a line parallel to AB.

Ex. **603**. Use the locus of Ex. 602 to solve the following problem. O is a point in the angle formed by two lines AB, AC. Through O draw a line, terminated by AB, AC, and bisected at O.

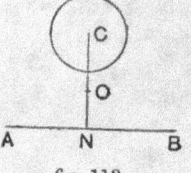

Ex. **604**. Draw a figure like fig. 113, making radius of circle 2 ins., CO=3 ins., CN=5 ins. Through O draw a line (or lines), terminated by AB and the circle, and bisected at O. (See Ex. 602.)

fig. 113.

Ex. **605**. A town X is 2 miles from a straight railway; but the two stations nearest to X are each 3 miles from X. Find the distance between the two stations.

CONSTRUCTION OF TRIANGLES, ETC. BY MEANS OF LOCI.

*Ex. **606**. Construct △ABC, given

   (i)   base BC=14 cm., height=9 cm., ∠B=65°. Measure AB.

   (ii) AB=59 mm., AC=88 mm., height AD=49 mm. (Draw height first.) Measure base BC.

   (iii) BC=4 in., ∠B=80°, median CN=4 in. Measure BA.

   (iv) base BC=12 cm., height AD=4 cm., median AL=5 cm. Measure AB, AC.

*Ex. **607**. Construct a right-angled triangle, given

   (i)   longest side=10 cm., another side=5 cm. Measure the smallest angle.

   (ii) side opposite right angle=4 in., another side=3 inches. Measure the third side.

*Ex. **608**. Construct a right-angled triangle ABC, given ∠A=90°, AB=7 cm., distance of A from BC=2·5 cm. Measure the smallest angle.

*Ex. **609**. Construct a triangle, given height=2 in., angles at the extremities of the base=40° and 60°. Find length of base.

*Ex. **610**. Construct a quadrilateral ABCD, given diagonal AC=9 cm., diagonal BD=10 cm., distances of B, D from AC 5 cm. and 4 cm. respectively, side AB=7 cm. Measure CD.

* Numerical values are given in these Exs., so that an accurate drawing may be made if desired; in cases in which technical skill is unimportant, freehand sketches will be sufficient.

*Ex. **611**. Construct a trapezium ABCD, given base AB=10 cm., height=4 cm., AD=4·5 cm., BC=4·2 cm.   Measure angles A and B. (There are 4 cases.)

*Ex. **612**. Construct a trapezium ABCD, given base AB=3·5 in., height=1·7 in., diagonals AC, BD=2·5, 3·5 ins. respectively.   Measure CD.

SYMMETRY.

¶Ex. **613**. Fold a piece of paper once; cut the folded sheet into any pattern you please; then open it out (see fig. 114).

The figure you obtain is said to be **symmetrical** about the line of folding. This line is called an **axis** of symmetry.

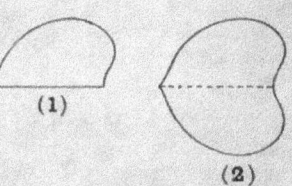

fig. 114.

¶Ex. **614**. Which of the following figures possess an axis of symmetry? (You may find that in some cases there is more than one axis.)   In each case make a sketch showing the axis (or axes), if there is symmetry. (i) isosceles △, (ii) equilateral △, (iii) square, (iv) rectangle, (v) parallelogram, (vi) rhombus, (vii) regular 5-gon, (viii) regular 6-gon, (ix) circle, (x) a semi-circle, (xi) a figure consisting of 2 unequal circles, (xii) a figure consisting of 2 equal circles.

Solids may have symmetry.   The human body is more or less symmetrical about a plane.   Consider the reflexion in a mirror of the interior of a room.   The objects in the room together with their reflexions form a symmetrical whole; the surface of the mirror is the **plane of symmetry**.

¶Ex. **615**. Give 4 instances of solids possessing planes of symmetry.

¶Ex. **616**. Fold a sheet of paper once.   Prick a number of holes through the double paper, forming any pattern.   On opening the paper you will find that the pin-holes have marked out a symmetrical figure.

Join **corresponding points** as in fig. 115.   Notice that when the figure was folded NP′ fitted on to NP.   This shows that NP′=NP.

* See note, p. 105.

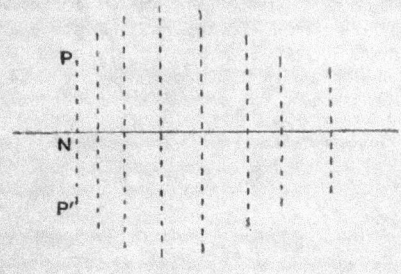

fig. 115.

This exercise shows that the line joining any pair of corresponding points, in a figure which is symmetrical about an axis, is bisected by and perpendicular to the axis of symmetry.

Corresponding points (e.g. P and P′ in fig. 115) are said to be **images** of one another in the axis of symmetry.

¶Ex. **617.** If a point P lies on the axis of symmetry, where is the image of P?

¶Ex. **618.** What is the image of a semicircle in its bounding diameter?

¶Ex. **619.** If a point moves along any curve, what is the locus of its image in a given line?

MISCELLANEOUS EXERCISES.

CONSTRUCTIONS.

Ex. **620.** A ship is sailing due N. at 8 miles an hour. At 3 o'clock a lighthouse is observed to be N.E. and after 90 minutes it is observed to bear $82\frac{1}{2}°$ E. of S. How far is the ship from the lighthouse at the second observation, and at what time (to the nearest minute) was the ship nearest to the lighthouse?

Ex. **621.** Is it possible to make a pavement consisting of equal equilateral triangles?

Is it possible to do so with equal regular figures of (i) 4, (ii) 5, (iii) 6, (iv) 7 sides?

Ex. 622. Show how to construct an isosceles triangle having given the base and the perpendicular from the vertex to the base.

†**Ex. 623.** Show how to construct an isosceles triangle on a given base, having its vertical angle equal to a given angle.   Give a proof.

†**Ex. 624.** Through one vertex of a given triangle draw a straight line cutting the opposite side, so that the perpendiculars upon the line from the other two vertices may be equal.   Give a proof.   [See Ex. 649.]

†**Ex. 625.** From a given point, outside a given straight line, draw a line making with the given line an angle equal to a given angle.   Give a proof. [Use parallels.]

†**Ex. 626.** Through a given point P draw a straight line to cut off equal parts from the arms of a given angle XOY.   Give a proof.   [Use parallels.]

Ex. 627. Draw a triangle ABC in which ∠B is less than ∠C.   Show how to find a point P in AB, such that PB=PC.

†**Ex. 628.** In the equal sides AB, AC of an isosceles triangle ABC, show how to find points X, Y such that BX=XY=YC.   Give a proof.

Ex. 629. A triangle ABC has ∠B=60°, BC=8 cm.; what is the least possible size for the side CA?   What is the greatest possible size for ∠C?

Ex. 630. Draw a triangle ABC and show how to find points P, Q in AB, AC such that PQ is parallel to the base BC and =⅓BC.
[Trisect the base and draw a parallel to one of the sides.]

†**Ex. 631.** In OX, OY, show how to find points A, B such that ∠OAB=3∠OBA.   Give a proof.
[What angle is equal to the sum of these angles?]

†**Ex. 632.** A and B are two fixed points in two unlimited parallel straight lines; show how to find points P and Q in these lines such that APBQ is a rhombus.   Give a proof.

†**Ex. 633.** Prove the following construction for bisecting the angle BAC :—With centre A describe two circles, one cutting AB, AC in D, E, and the other cutting them in F, G respectively; join DG, EF, intersecting in H; join AH.

†**Ex. 634.** A, B are two points on opposite sides of a straight line CD; in CD show how to find a point P such that ∠APC=∠BPD.   Give a proof.

†Ex. **635.** A, B are two points on the same side of a straight line CD; in CD show how to find a point P such that ∠APC=∠BPD. Give a proof.
[From A draw AN perpendicular to CD and produce it to A′ so that NA′=NA; if P is any point in CD, ∠ˢ APN and A′PN can be proved equal.]

†Ex. **636.** A, B are two points on opposite sides of a straight line CD; in CD show how to find a point P so that ∠APC=∠BPC. Give a proof.

†Ex. **637.** Show how to construct a rhombus PQRS having its diagonal PR in a given straight line and its sides PQ, QR, RS passing through three given points L, M, N respectively. Give a proof.

Ex. **638.** Show how to describe a rhombus having two of its sides along the sides AB, AC of a given triangle ABC and one vertex in the base of the triangle.

†Ex. **639.** Show how to draw a straight line equal and parallel to a given straight line and having its ends on two given straight lines. Give a proof.

¶Ex. **640.** **To trisect a given angle.**

Much time was devoted to this famous problem by the Greeks and the geometers of the Middle Ages; it has now been shown that it is impossible with only the aid of a pair of compasses and a straight edge (ungraduated).

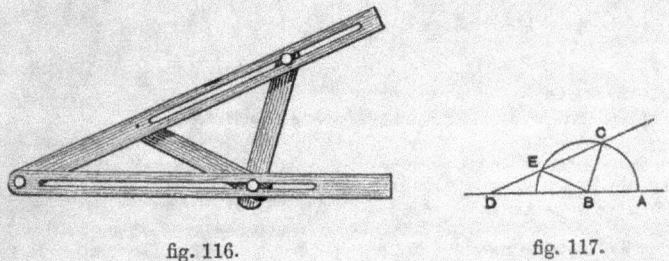

fig. 116.   fig. 117.

Fig. 116 shows a simple form of trisector; the instrument is opened until the angle between the rods corresponding to BA and BC can be made to coincide with the given angle; then the angle between the long rods (corresponding to ∠ADC) is one-third of the given angle. To prove this, take DE = the radius of the circle (in fig. 117), and show that ∠ADC=⅓∠ABC.

With a ruler, marked on its edge in two places, and a pair of compasses, it is possible to trisect an angle as follows:—

Let ABC be the angle. With B as centre and radius = the distance between the two marks describe a circle cutting BC at C; place the ruler so that its edge passes through C and has one mark on AB produced, the other on the circle (this must be done by trial, a pin stuck through the paper at C will help); rule the line DEC, then $\angle D = \frac{1}{3} \angle ABC$.

THEOREMS.

¶ Ex. **641**. If one side of a triangle is *double* another, is the angle opposite the former double the angle opposite the latter?

In order to answer this question, take the following instances:

(1) Consider a triangle whose angles are 45°, 45°, 90°.

(2) Consider a triangle whose angles are 30°, 60°, 90°.

(3) Draw △ABC in which AB = 8·2 cms., BC = 4·1 cms., CA = 6 cms. Measure the angles. Is C double A?

(4) Draw △ABC in which ∠A = 82°, ∠B = 41°, BC = 3″. Measure the remaining sides. Is BC double CA?

(5) Prove that in △ABC whose angles are 30°, 60°, 90°, the *longest* side AB is double the *shortest* BC.

[Make ∠CAD = 30° and produce BC to meet AD in D.

How many degrees are there in ∠D?

What kind of a triangle is ABD?]

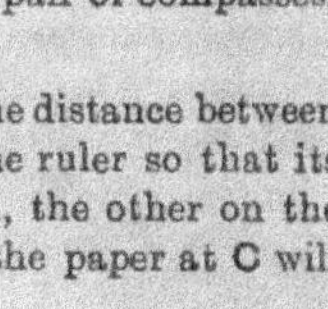

fig. 118.

Ex. **642**. The gable end of a house is in the form of a pentagon, of which the three angles at the ridge and eaves are equal to each other : show that each of these angles is equal to twice the angle of an equilateral triangle.

†Ex. **643**. If on the sides of an equilateral triangle three other equilateral triangles are described, show that the complete figure thus formed will be (i) a triangle, (ii) equilateral.

†Ex. **644**. Two isosceles triangles are on the same base: prove that the straight line joining their vertices bisects the base at right angles.

†Ex. **645**. Two triangles ABC, DCB stand on the same base BC and on the same side of it; prove that AD is parallel to BC if AB = DC and AC = DB.

†Ex. **646.** If straight lines are drawn from a point perpendicular to the arms of an angle, the angle between those straight lines is equal or supplementary to the given angle.

†Ex. **647.** If points X, Y, Z are taken in the sides BC, CA, AB of an equilateral triangle, such that $\angle BAX = \angle CBY = \angle ACZ$, prove that, unless AX, BY, CZ pass through one point, they form another equilateral triangle.

Ex. **648.** If points X, Y, Z are taken in the sides BC, CA, AB of any triangle, such that $\angle BAX = \angle CBY = \angle ACZ$, prove that, unless AX, BY, CZ pass through one point, they form a triangle whose angles are equal to the angles of the triangle ABC.

†Ex. **649.** The extremities of a given straight line are equidistant from any straight line drawn through its middle point.

†Ex. **650.** In the diagonal AC of a parallelogram ABCD points P, Q are taken such that AP = CQ; prove that BPDQ is a parallelogram.

†Ex. **651.** ABCD, ABXY are two parallelograms on the same base and on the same side of it. Prove that CDYX is a parallelogram.

†Ex. **652.** The diagonal AC of a parallelogram ABCD is produced to E, so that CE = CA; through E, EF is drawn parallel to CB to meet DC produced in F. Prove that ABFC is a parallelogram.

†Ex. **653.** E, F, G, H are points in the sides AB, BC, CD, DA respectively of a parallelogram ABCD, such that AH = CF and AE = CG; show that EFGH is a parallelogram.

†Ex. **654.** C is the mid-point of AB; from A, B, C perpendiculars AX, BY, CZ are drawn to a given straight line. Prove that, if A and B are both on the same side of the line, AX + BY = 2CZ.

What relation is there between AX, BY, CZ when A and B are on opposite sides of the line?

†Ex. **655.** If the bisectors of the base angles of an isosceles triangle ABC meet the opposite sides in E and F, EF is parallel to the base of the triangle.

†Ex. **656.** In a quadrilateral ABCD, AB = CD and $\angle B = \angle C$; prove that AD is parallel to BC.

†Ex. **657.** Prove that the diagonals of an isosceles trapezium are equal.

†Ex. **658.** Two rods OA, OB, are jointed at O. P, Q are the middle points of OA, OB respectively, and two rods PR, QR are fitted so as to form a parallelogram OPRQ jointed at the vertices. Prove that A, R, B are in a straight line however the mechanism may move.

†Ex. **659.** AB, AC are equal straight lines. D and E are points, D in AB, E in AC produced, such that DB=CE; prove that DE is bisected by BC.

†Ex. **660.** From a point C within the acute angle formed by two lines OA, OB a line is drawn parallel to OA to meet OB in B. The circle whose centre is C and radius CB cuts OB again in D; DC produced meets OA in A. Prove that OA is equal to AD.

†Ex. **661.** ABCD is a quadrilateral, such that $\angle A = \angle B$ and $\angle C = \angle D$; prove that AD=BC.

†Ex. **662.** The figure formed by joining the mid-points of the sides of a rectangle is a rhombus.

†Ex. **663.** The medians BE, CF of a triangle ABC intersect at G; GB, GC are bisected at H, K respectively. Prove that HKEF is a parallelogram. Hence prove that G is a point of trisection of BE and CF.

†Ex. **664.** The diagonal AC of a parallelogram ABCD is produced to E, so that CE=CA; through E and B, EF, BF are drawn parallel to CB, AC respectively. Prove that ABFC is a parallelogram.

†Ex. **665.** T, V are the mid-points of the opposite sides PQ, RS of a parallelogram PQRS. Prove that ST, QV trisect PR.

†Ex. **666.** Any straight line drawn from the vertex to the base of a triangle is bisected by the line joining the mid-points of the sides.

†Ex. **667.** The sides AB, AC of a triangle ABC are produced to X, Y respectively, so that BX=CY=BC; BY, CX intersect at Z. Prove that $\angle BZX + \frac{1}{2} \angle BAC = 90°$.

†Ex. **668.** ABCD is a parallelogram and AD=2AB; AB is produced both ways to E, F so that EA=AB=BF. Prove that CE, DF intersect at right angles.

†Ex. **669.** In fig. 109, if AB, CD are parallel and AC=CE and BD=DF, prove that EF is parallel to CD.

†Ex. **670.** In a right-angled triangle, the distance of the vertex from the mid-point of the hypotenuse is equal to half the hypotenuse.

[Join the mid-point of the hypotenuse to the mid-point of one of the sides.]

†Ex. **671.** Given in position the right angle of a right-angled triangle and the length of the hypotenuse, find the locus of the mid-point of the hypotenuse. (See Ex. 670.)

†Ex. **672**. ABCD is a square; from A lines are drawn to the mid-points of BC, CD; from C lines are drawn to the mid-points of DA, AB. Prove that these lines enclose a rhombus.

†Ex. **673**. ABC is an equilateral triangle and D is any point in AB; on the side of AD remote from C an equilateral triangle ADE is described; prove that BE=CD.

†Ex. **674**. In a triangle ABC, BE and CF are drawn to cut the opposite sides in E and F; prove that BE and CF cannot bisect one another.

†Ex. **675**. ABC is an acute-angled triangle, whose least side is BC. With B as centre, and BC as radius, a circle is drawn cutting AB, AC at D, E respectively. Show that, if AD=DE, ∠ABC=2∠BAC.

†Ex. **676**. ABC is an isosceles triangle (AB=AC); a straight line is drawn cutting AB, BC, and AC produced in D, E, F respectively. Prove that, if DE=EF, BD=CF.

†Ex. **677**. If two triangles have two sides of the one equal to two sides of the other, each to each, and the angles opposite to two equal sides equal, the angles opposite the other equal sides are either equal or supplementary; and in the former case the triangles are congruent.

†Ex. **678**. A quadrilateral ABCD, that has AB=AD and BC=DC, is called a **kite**. Use Th. I. 25 to prove that the diagonals of a kite are at right angles.

†Ex. **679**. If two circles cut at P, Q, use I. 25 to prove that the line joining their centres bisects PQ at right angles.

BOOK II.

AREA.

Area of rectangle. In the course of your arithmetic and laboratory work you have learnt the following rule:

To find the number of square units in the area of a rectangle, multiply together the numbers of units in the length and breadth of the rectangle.

¶Ex. 680. Draw a figure to show that if the side of one square is 3 times the side of another square, the area of the one square is 9 times the area of the other. (*Freehand.*)

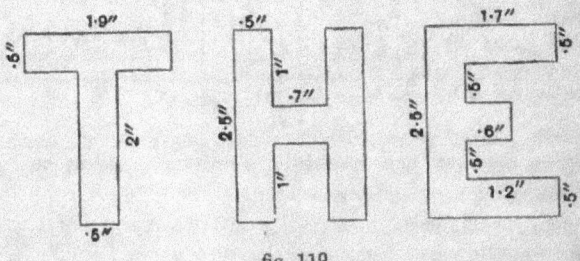

fig. 119.

Ex. 681. Make freehand sketches of the given figures (fig. 119). In each case find the area.

Ex. 682. Find the other dimension of a rectangle, given that the area = 140 sq. ft., one dimension = 35 ft. Now state the rule for finding the other dimension, given the area and one dimension.

Area of right-angled triangle. By drawing a diagonal of a rectangle we divide the rectangle into two equal right-angled triangles. Hence the area of a right-angled triangle may be found by regarding it as half a certain rectangle.

Ex. 683. Find the areas of right-angled triangles in which the sides containing the right angle are (i) 2″, 3″, (ii) 6·5 cm., 4·4 cm.

Area of any rectilinear figure (on squared paper). With the aid of rectangles and right-angled triangles we can find the area of any figure contained by straight lines (i.e. any **rectilinear** figure). This way is especially convenient when one side of the figure runs along a line of the squared paper.

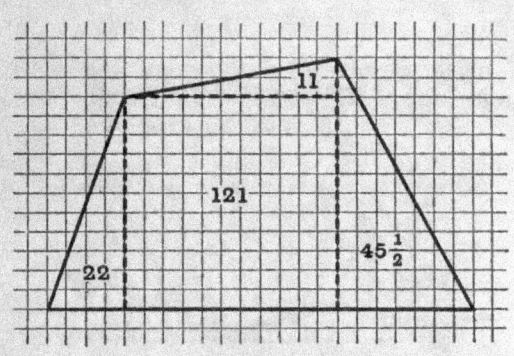

fig. 120.

Fig. 120 shows how a 4-sided figure may be divided up into rectangles and right-angled triangles; the number inside each rectangle and triangle indicates the number of squares it contains; and the complete area is $199\frac{1}{2}$ or 199·5 squares.

Ex. **684.** Find the area (in squares of your paper) of each of the following figures by dividing up the figures into rectangles and right-angled triangles:

   (i)   (2, 1), (11, 1), (8, 6), (2, 6).

   (ii)   (1, 2), (1, 10), (6, 13), (6, 2).

   (iii) (5, 0), (3, 4), (–5, 4), (–6, 0).

   (iv) (0, 6), (–3, 2), (–3, –2), (0, –3).

   (v)   (0, 0), (1, 4), (6, 0).

   (vi) (1, 4), (6, 3), (1, –3).

  (vii) (–4, –3), (–3, 3), (5, 6), (10, –3).

 (viii) (3, 5), (–3, 2), (–5, –3), (3, –7).

   (ix) (3, 0), (0, 6), (–3, 0), (0, –6).

   (x)   (2, 5), (5, 2), (5, –2), (2, –5), (–2, –5), (–5, –2), (–5, 2), (–2, 5).

8—2

If there is no side of the figure which coincides with a line of the paper (ABCD in fig. 121), it is generally convenient to draw lines *outside* the figure, parallel to the axes, thus making up a rectangle (PQRS); the area required can then be found by subtracting a certain number of right-angled triangles from the rectangle.

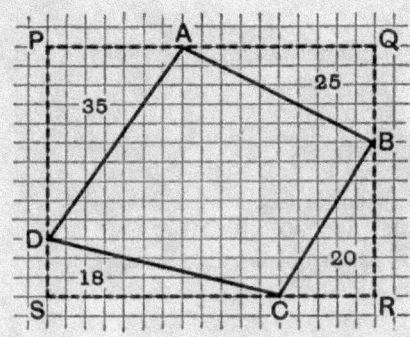

fig. 12.

Thus in fig. 121

$$ABCD = PQRS - AQB - BRC - CSD - DPA$$
$$= 221 - 25 - 20 - 18 - 35$$
$$= 123.$$

Ex. 685. Find the areas of the following figures:

(i)   (1, 1), (16, 5), (9, 14).

(ii)   (6, 3), (12, 9), (3, 11).

(iii) (10, −20), (20, −24), (12, 4).

(iv) (0, 0), (9, −1), (7, 6), (2, 5).

(v)   (1, 0), (6, 1), (5, 6), (0, 5).

(vi) (3, 0), (7, 3), (4, 7), (0, 4).

(vii) (4, 0), (10, 4), (6, 10), (0, 6).

(viii) (5, 0), (0, 5), (−5, 0), (0, −5).

DEF. Any side of a **parallelogram** may be taken as the **base**. The perpendicular distance between the base and the opposite (parallel) side is called the **height**, or **altitude**.

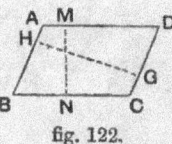

fig. 122.

Thus in fig. 122 if BC be taken as base, MN (which may be drawn from any point of the base) is the height (or altitude). If AB be taken as base, GH is the height.

¶Ex. **686.** In fig. 122 what is the height if CD be taken as base? if AD be taken?

Area of parallelogram. Take a sheet of paper (a rectangle) and call the corners P, B, C, Q; BC being one of the longer sides (fig. 123). Mark a point A on the side PQ. Join BA, and cut (or tear) off the right-angled triangle PBA. You now have two pieces of paper; you will find that you can fit them together to make a parallelogram (ABCD in fig. 123).

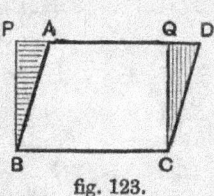

fig. 123.

Notice (i) that the rectangle you had at first and the parallelogram you have now made, are composed of the same paper, and *therefore have the same area.*

(ii) that the rectangle and the parallelogram are on the same base BC, and both lie between the same pair of parallel lines BC and PAQD. Or, we may say that they have *the same height.*

DEF. Figures which are equal in area are said to be **equivalent.**

Notice that congruent figures are necessarily equivalent; but that equivalent figures are not necessarily congruent.

THEOREM 1.

Parallelograms on the same base and between the same parallels (or, of the same altitude) are equivalent.

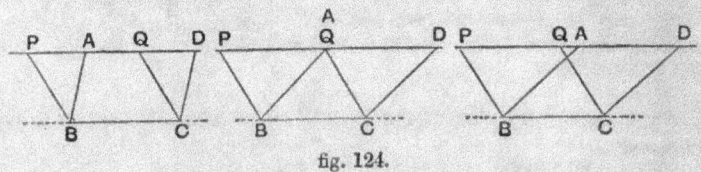

fig. 124.

Data   ABCD, PBCQ are $\parallel^{\text{ograms}}$ on the same base BC, and between the same parallels BC, PD.

To prove that   ABCD and PBCQ are equivalent.

Proof   In the △s PBA, QCD,

∠ BAP = corresp. ∠ CDQ (∵ BA, CD are ∥),   I. 5.

∠ BPA = corresp. ∠ CQD (∵ BP, CQ are ∥),   I. 5.

BA = CD (opp. sides of $\parallel^{\text{ogram}}$ ABCD),   I. 22.

∴ the triangles are congruent.   I. 11.

Now if △ PBA is subtracted from figure PBCD, $\parallel^{\text{ogram}}$ BD is left; and if △ QCD is subtracted from figure PBCD, $\parallel^{\text{ogram}}$ BQ is left.

Hence the $\parallel^{\text{ograms}}$ are equivalent

Q. E. D.

Cor. 1. Parallelograms on equal bases and of the same altitude are equivalent.

(For they can be so placed as to be on the same base and between the same parallels.)

Cor. 2. **The area of a parallelogram is measured by the product of the base and the altitude.**

(For the $\parallel^{\text{ogram}}$ is equivalent to a rectangle on the same base and of the same altitude, whose area = base × altitude.)

Ex. 687.   Find the area of a parallelogram of sides 2 ins. and 3 ins. and of angle 30°.

Ex. 688.   Draw a rectangle on base 12 cm. and of altitude 10 cm.; on the same base construct an equivalent parallelogram of angle 30°; and measure its longer diagonal.

Ex. 689.   Show how to construct a parallelogram equivalent to a given rectangle, on the same base and having one of its angles equal to a given angle (without using protractor).

Ex. 690.   Draw a rectangle of base 4 ins., and height 3 ins.: on the same base make an equivalent parallelogram with a pair of sides of 5 ins. Measure the angle between the base and the shorter diagonal.

Ex. 691.   Show how to construct on the same base as a given rectangle an equivalent parallelogram having its other side equal to a given straight line (without using scale).   Is this always possible?

Ex. 692.   Transform a rectangle of base 4·5 cm. and height 2·9 cm. into an equivalent parallelogram having a diagonal of 8·4 cm.   Measure the angle between the base and that diagonal.

Ex. 693.   Transform a parallelogram of sides 2 and 1 ins. and angle 80° into an equivalent parallelogram of sides 2 and 2·5 ins.   Measure acute angle of the latter.

Ex. 694.   Transform a parallelogram of base 2·34 ins., height 2·56 ins. and angle 67° into an equivalent parallelogram on the same base with angle 60°.   Measure the other side of the latter.

Ex. 695.   Make parallelogram ABCD, with AB = 2·5 ins., AD = 3 ins., angle A = 60°.   Transform this into an equivalent parallelogram with sides of 2 ins. and 4 ins. ; measure the acute angle of the latter.

(First, keeping the same base AB, make equivalent parallelogram ABEF having AE = 4 ins.   Next, taking AE for base, construct an equivalent parallelogram with sides 2 and 4 ins.)

Ex. 696.   Show how to make a parallelogram equivalent to a given rectangle, having its sides equal to two given lines.   Is this always possible?

Ex. 697.   What is the locus of the intersection of the diagonals of a parallelogram whose base is fixed and area constant?

Area of Triangle.

Def.   Any side of a triangle may be taken as base. The line drawn perpendicular to the base from the opposite vertex is called the **height, or altitude.**

There will be three different altitudes according to the side which is taken as base.

¶Ex. **698.**   Draw an acute-angled triangle and draw the three altitudes. (*Freehand.*)

¶Ex. **699.**   Repeat Ex. 698 for a right-angled triangle   (*Freehand.*)

¶Ex. **700.**   Repeat Ex. 698 for an obtuse-angled triangle   (*Freehand*)

¶Ex. **701.**   In what case are two of the altitudes of a triangle equal?

Theorem 2.

Triangles on the same base and between the same parallels (or, of the same altitude) are equivalent.

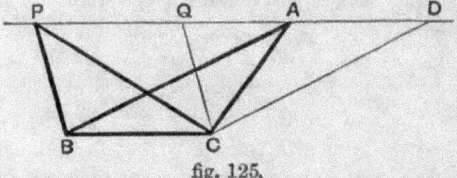

fig. 125.

Data   ABC, PBC are △s on the same base BC, and between the same parallels BC, PA.

To prove that   ABC, PBC are equivalent.

Construction   Complete the ‖ᵒᵍʳᵃᵐˢ ABCD, PBCQ by drawing CD, CQ ‖ to BA, BP respectively, to meet PA (produced if necessary) in D, Q.

Then △ ABC = ½ ‖ᵒᵍʳᵃᵐ ABCD,   I. 22 (3).

and △ PBC = ½ ‖ᵒᵍʳᵃᵐ PBCQ.   I. 22 (3).

But ‖ᵒᵍʳᵃᵐˢ ABCD, PBCQ are equivalent, being on the same base and between the same parallels.   II. 1.

∴   △ ABC = △ PBC.

Q. E. D.

COR. 1. Triangles on equal bases and of the same altitude are equivalent.

(For they can be so placed as to be on the same base and between the same parallels.)

COR. 2. **The area of a triangle is measured by half the product of the base and the altitude.**

†Ex. 702. **Prove Cor. 2.**

†Ex. 703. Prove that the area of a right-angled triangle is half the product of the sides which contain the right angle.

¶Ex. 704. A triangle ABC has AB = 3 ins. and BC = 4 ins.; trace the changes in its area as ∠ABC increases from 0° to 180°.

¶Ex. 705. A parallelogram PQRS has PQ = 10 cm. and QR = 7 cm.; trace the changes in its area as ∠PQR increases from 0° to 180°.

Since any one of the three sides may be taken for base, there are three different ways of forming the product of a base and the corresponding altitude. Thus the area may be calculated in three different ways; and of course, theoretically, the result is the same in each case. Practically, none of the measurements will be quite exact, and the results will generally differ slightly. To get the best possible value for the area **take the average of the three results.**

Ex. 706. Find, to three significant figures, the areas of the following triangles, taking the average of three results in each case:

    (i)   sides 3, 4, 4·5 ins.

    (ii)   sides 6, 8, 9 cm.

    (iii) sides 3, 4, 5 ins.

    (iv) sides 2, 3, 4·5 ins.

    (v)   sides 3, 4 ins., included ∠120°.

    (vi) BC = 7·2 cm., ∠B = 20°, ∠C = 40°.

Ex. 707. (On inch paper.) The vertices of a triangle are the points (2, 0), (−1, 2), (−2, −2). Find the area (i) by measuring sides and altitudes, (ii) as on p. 116.

Ex. **708.** Find the surface (i.e. the sum of the areas of all the faces):

(i)   of a tetrahedron (three-sided pyramid) each of whose edges is 4 inches.

(ii)   of a square pyramid whose base edges are each 2 inches and whose slant edges are each 2·5 inches.

(iii)   of a regular 3-sided prism whose base is an equilateral triangle of side 2 inches and whose height is 3·5 inches.

Ex. **709.** Find the combined area of the walls and roof of the house in fig. 70; take width of house = 8 yds., depth (front to back) = 4 yds., height of front wall = 6 yds., height of roof-ridge above ground = $7\frac{1}{2}$ yds.   Neglect doors and windows.

†Ex. **710.** Prove that the area of a rhombus is half the product of its diagonals.

†Ex. **711.** **D is the mid-point of the base BC of a triangle ABC; prove that triangles ABD, ACD are equivalent.**

†Ex. **712.** ABCD is a parallelogram; P, Q the mid-points of AB, AD. Prove that $\triangle APQ = \frac{1}{8}$ of ABCD.   (Join PD, BD.)

†Ex. **713.** The base BC of $\triangle$ABC is divided at D so that $BD = \frac{1}{3}BC$; prove that $\triangle ABD = \frac{1}{3} \triangle ABC$.

†Ex. **714.** The base BC of $\triangle$ABC is divided at D so that $BD = \frac{2}{5}BC$; prove that $\triangle ABD = \frac{2}{3} \triangle ACD$.

†Ex. **715.** **The ratio of the areas of triangles of the same height is equal to the ratio of their bases.**

†Ex. **716.** The ratio of the areas of triangles on the same base is equal to the ratio of their heights.

†Ex. **717.** ABCD is a quadrilateral and the diagonal AC bisects the diagonal BD. Prove that AC divides the quadrilateral into equivalent triangles (fig. 126).

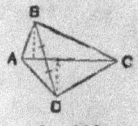

fig. 126.

†Ex. **718.** E is the mid-point of the diagonal AC of a quadrilateral ABCD. Prove that the quadrilaterals ABED, CBED are equivalent.

†Ex. **719.** E is a point on the median AD of $\triangle$ABC; prove that $\triangle ABE = \triangle ACE$.

†Ex. **720.** D is a point on the base BC of △ ABC; E is the mid-point of AD; prove that △ EBC = ½ △ ABC.

†Ex. **721.** Divide a triangle into 4 equivalent triangles. (*Freehand.*)

Ex. **722.** The base of a triangle is a fixed line of length 3 inches, and the vertex moves so that the area of the triangle is always 6 sq. ins. What is the altitude? What is the locus of the vertex?

†Ex. **723.** Prove that the locus of the vertex of a triangle of fixed base and constant area is a pair of straight lines parallel to the base.

Ex. **724.** Draw a scalene triangle, and transform it into an equivalent isosceles triangle on the same base. (Keep the base fixed; where must the vertex be in order that the triangle may be isosceles? Where must the vertex be in order that the triangle may be equivalent to the given triangle?) (*Freehand.*)

Ex. **725.** Show how to transform a given triangle

(i) into an equivalent right-angled triangle.

(ii) into an equivalent triangle on the same base, having one side of 2 inches. Is this always possible?

(iii) into an equivalent triangle having one angle = a given angle.

(iv) into an equivalent right-angled triangle with one of the sides about the right angle equal to 5 cm. (First make one side 5 cm.; then take this as base and make the triangle right-angled.)

(v) into an equivalent isosceles triangle with base equal to a given line.

Ex. **726.** Transform an equilateral triangle of side 3 ins. into an equivalent triangle with a side of 4 ins., and an angle of 60° adjacent to that side. Measure the other side adjacent to the 60° angle.

Ex. **727.** Transform a given triangle into an equivalent triangle with its vertex (i) on a given line, (ii) one inch from a given line, (iii) one inch from a given point, (iv) equidistant from two given intersecting lines.

†Ex. **728.** In fig. 127 PA is parallel to BC. Prove that △ POB = △ AOC.

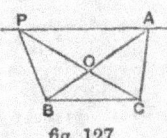

fig. 127.

†Ex. **729.** A line parallel to the base BC of △ ABC cuts the sides AB, AC in D, E respectively. Prove that △ ABE = △ ACD.

†Ex. **730**.   F is any point on the base BC of △ ABC : E is the mid-point of BC.   ED is drawn parallel to AF.   Prove that △ DFC=½△ ABC.   (See fig. 128.)

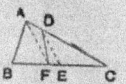

fig. 128.

†Ex. **731**.   Draw a line through a given point of a side of a triangle to bisect the area of the triangle.   (See Ex. 730.)

¶Ex. **732**.   Transform a given quadrilateral ABCD into an equivalent quadrilateral ABCD′, so that the three vertices A, B, C may be unchanged, and ∠ BAD′=170°.

¶Ex. **733**.   Repeat Ex. 732, making ∠ BAD′=180°.   What kind of figure is produced?

Area of any rectilinear figure.   This may be determined in various ways.

METHOD I.   By dividing up the figure into triangles.

METHOD II.   Perhaps the most convenient method is that of constructing a single triangle equivalent to the given figure, as follows:

To construct a triangle equivalent to a given quadrilateral ABCD.

Construction   Join CA.   Through D draw
    DD′ ∥ CA, meeting BA produced in D′.
        Join CD′.
        Then △ BCD′ = quadrilateral ABCD.

Proof   △ ACD′ = △ ACD.   (Why?)
        Add to each △ ACB.

            ∴ △ BCD′ = quadrilateral ABCD.

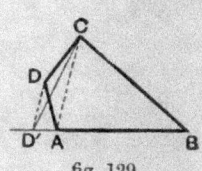

fig. 129.

In a similar way a pentagon may be reduced, first to an equivalent quadrilateral and then to an equivalent triangle : and so for figures of more sides. The area of the triangle can then be found as already explained. A convenient method of dealing with the pentagon is shown in fig. 130.

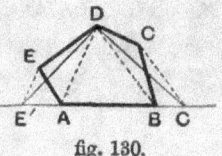

fig. 130.

†Ex. **734.** Explain the construction of fig. 130, and prove that

$$\triangle \ C'DE' = \text{figure ABCDE.}$$

†Ex. **735.** Given a quadrilateral ABCD, construct an equivalent triangle on base AB having ∠A in common with the quadrilateral. (*Freehand.*)

Ex. **736.** Find the area of a quadrilateral ABCD, when

(i) DA=1 in., ∠A=100°, AB=2·3 ins., ∠B=64°, BC=1·5 ins.

(ii) AB=5·7 cm., BC=5·2 cm., CD=1·7 cm., DA=3·9 cm., ∠A=76°.

Ex. **737.** Find the area of a pentagon ABCDE, given AB=6·5 cm., BC=2·4 cm., CD=DE=4 cm., EA=2·5 cm., ∠A=80°, ∠B=133°.

Ex. **738.** Find the area of a **trapezium** ABCD (fig. 131), given AB=3 ins., height=2 ins., ∠A=70°, ∠B=50°. (Divide into 2 △s, and notice that their heights DE, BF are equal.)

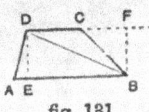

fig. 131.

†Ex. **739.** In fig. 132 E is the mid-point of BC, PQ is ∥ to AD. Prove that

trapezium ABCD = ∥ogram APQD.

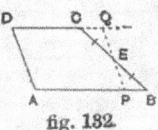

fig. 132.

†Ex. **740. Prove that the area of a trapezium is equal to half the product of the height and the sum of the two parallel sides.**

THEOREM 3.

Equivalent triangles which have equal bases in the same straight line, and are on the same side of it, are between the same parallels.

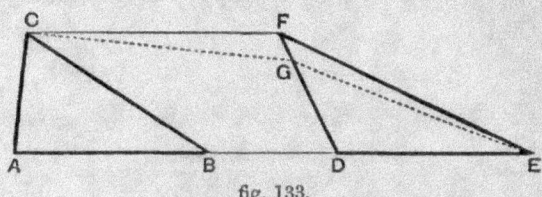

fig. 133.

Data   ABC DEF are equivalent triangles on equal bases AB, DE, these being in a straight line, and C and F being on the same side of AE.

To prove that   CF is parallel to AE.

Construction   Join CF.

If possible, draw a line CG ‖ to AE, distinct from CF, meeting FD (produced if necessary) in G.   Join EG.

Proof   Since AB = DE, and CG is ‖ to AE,

∴ △ ABC = △ DEG.   II. 2.

But △ ABC = △ DEF,   *Data*

∴ △ DEF = △ DEG,

∴ F coincides with G, and CF with CG,

∴ CF is ‖ to AE.

Q. E. D.

COR. 1.   Equivalent triangles on the same or equal bases are of the same altitude.

COR. 2.   Equivalent triangles on the same base and on the same side of it are between the same parallels.

NOTE.   The method of proof adopted in the above theorem is called **reductio ad absurdum.**

†Ex. **741.**   Give another proof of Cor. 1.

Ex. **742.** What is the converse of the above Theorem?

†Ex. **743.** D, E are the mid-points of the sides AB, AC of a triangle ABC; prove that DE is parallel to BC. (Join DC, EB.)

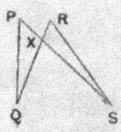

fig. 134.

†Ex. **744.** In fig. 134 △ PXQ = △ RXS; prove that PR is parallel to QS.

†Ex. **745.** In fig. 135 △ AEB = △ ADC; prove that DE is parallel to BC.

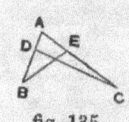

fig. 135.

THEOREM 4.

If a triangle and a parallelogram stand on the same base and between the same parallels, the area of the triangle is half that of the parallelogram.

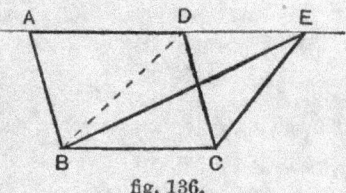

fig. 136.

Data △ EBC and ||^{ogram} ABCD stand on the same base BC and between the same parallels BC, AE.

To prove that △ EBC = ½ ||^{ogram} ABCD.

Construction Join BD.

Proof Since AE is || to BC,

∴ △ EBC = △ DBC, II. 2.

and △ DBC = ½ ||^{ogram} ABCD, I. 22.

∴ △ EBC = ½ ||^{ogram} ABCD.

Q. E. D.

* This theorem does not appear in the schedules of the Cambridge Previous Examination and Oxford Responsions.

†Ex. **746.** Show how to construct a rectangle equal to a given triangle. Give a proof.

†Ex. **747.** F, E are the mid-points of the sides AD, BC of a parallelogram ABCD; P is any point in FE.   Prove that △APB = ¼ABCD.

†Ex. **748.** P, Q are any points upon adjacent sides AB, BC of a parallelogram ABCD; prove that △CDP = △ADQ.

†Ex. **749.** AB, CD are parallel sides of a trapezium ABCD; E is the mid-point of AD; prove that △BEC = ½ trapezium.   (Through E draw a line parallel to BC.)

†Ex. **750.** O is a point inside a parallelogram ABCD; prove that

$$\triangle OAB + \triangle OCD = \tfrac{1}{2} ABCD.$$

THE THEOREM OF PYTHAGORAS.

Fig. 137 represents an isosceles right-angled triangle with
squares described upon each of the
sides.   The dotted lines divide up the
squares into right-angled triangles,
each of which is obviously equal to the
original triangle.   This sub-division
shows that the square on the hypotenuse of the above right-angled triangle is equal to the sum of the
squares on the sides containing the
right angle.   (A tiled pavement often
shows this fact very clearly.)

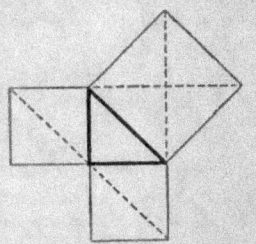

fig. 137.

¶Ex. **751.** Construct a right-angled triangle
with sides of 3 cm. and 4 cm. containing the right
angle.   Measure the length of the hypotenuse.
What relation is there between the square on the
hypotenuse and the squares on the other two
sides?   See fig. 138.

fig. 138.

Ex. 752. Draw a good-sized scalene right-angled triangle ABC, right-angled at A. Measure the three sides and calculate the areas of the squares upon them. Add together the areas of the two smaller squares, and arrange your results like this—

AB=...cm.,　sq. on AB=...sq. cm.,
AC=...cm.,　sq. on AC=...sq. cm.,
sum of sqq. on AB, AC=...sq. cm.,
BC=...cm.,　sq. on BC=...sq. cm.

The above exercises lead up to the fact that

"In a right-angled triangle the square described on the hypotenuse is equal to the sum of the squares on the other two sides."

This famous theorem was discovered by Pythagoras (B.C. 570 —500). Before proving it, the pupil may try the following experiment.

Ex. 753. Draw (on paper or, better, on thin cardboard) a right-angled triangle and the squares on the three sides (see fig. 139). Choose one of the two smaller squares and cut it up in the following manner. First find the centre of the square by drawing the diagonals. Then, through the centre, make a cut across the square parallel to BC, the hypotenuse, and a second cut perpendicular to BC. It will be found that the four pieces of this square together with the other small square exactly make up the square on the hypotenuse.

fig. 139.

(Perigal's dissection.)

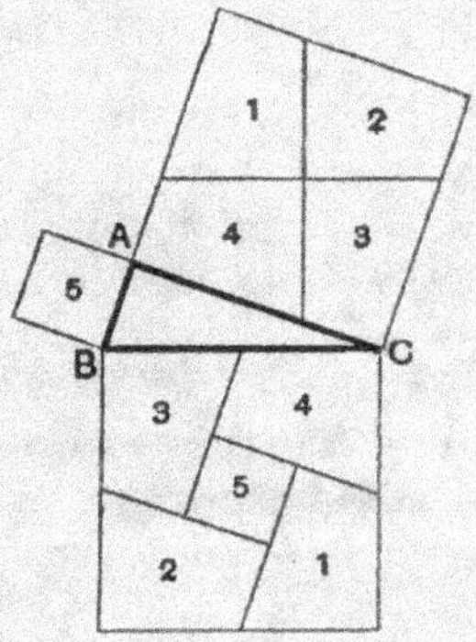

The following exercises lead up to the *method of proof* adopted for the theorem of Pythagoras.

†**Ex. 754.** On two of the sides AB, BC of *any* triangle ABC are described squares ABFG, BCED (as in fig. 140); prove that triangles BCF, BDA are congruent; and that CF=DA.

†**Ex. 755.** On the sides of any triangle ABC are described equilateral triangles BCD, CAE, ABF, their vertices pointing outwards. Prove that AD=BE=CF.

THEOREM 5.

[THE THEOREM OF PYTHAGORAS.]

In a right-angled triangle, the square on the hypotenuse is equal to the sum of the squares on the sides containing the right angle.

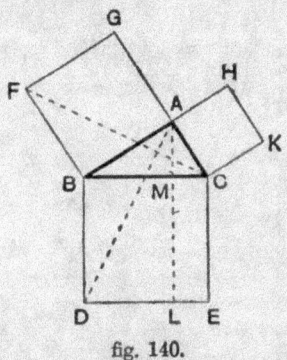

fig. 140.

Data   ABC is a triangle, right-angled at A.

The figures BE, CH, AF are squares described upon BC, CA, AB respectively.

To prove that   sq. BE = sq. CH + sq. AF.

Construction   Through A draw AL ‖ to BD (or CE).
Join CF, AD.

Proof

△ABD ≡ △FBC

rt. ∠ CBD = rt. ∠ FBA,
add to each ∠ ABC,
∴ ∠ ABD = ∠ FBC.
Hence, in △s ABD, FBC
⎰ ∠ ABD = ∠ FBC,
⎨   AB = FB (sides of a square),
⎱   BD = BC (,,   ,,   ,,),
∴ △ ABD ≡ △ FBC.   I. 10.

Since each of the angles BAC, BAG is a right angle,

∴ CAG is a st. line,   I. 2.

and this line is ∥ to BF.

$\triangle$ FBC = $\frac{1}{2}$ sq. AF

$\triangle$ ABD = $\frac{1}{2}$ rect. BL

∴ $\triangle$ FBC and sq. AF are on the same base BF, and between the same parallels BF, CG,

∴ $\triangle$ FBC = $\frac{1}{2}$ sq. AF.   II. 4.

Again $\triangle$ ABD and rect. BL are on the same base BD and between the same parallels BD, AL,

∴ $\triangle$ ABD = $\frac{1}{2}$ rect. BL.   II. 4.

But $\triangle$ FBC $\equiv$ $\triangle$ ABD.   *Proved*

∴ sq. AF = rect. BL.

sq. AF = rect. BL

sq. CH = rect. CL

∴ sq. AF + sq. CH

= sq. BE

In a similar way, by joining BK, AE, it may be shown that

sq. CH = rect. CL.

Hence

sq. AF + sq. CH = rect. BL + rect. CL

= sq. BE.

Q. E. D.

†Ex. **756**. Alternative proofs of Pythagoras' theorem are suggested (i) by fig. 141, (ii) by fig. 142. Write out these proofs.

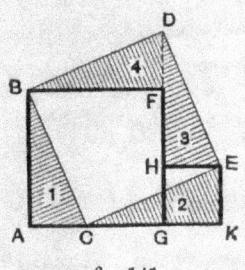
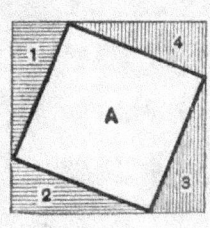
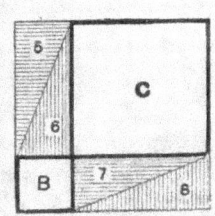

fig. 141.   fig. 142.

9—2

Ex. **757**. What is the square on the hypotenuse of a right-angled triangle if the sides containing the right angle are 6 cm. and 8 cm.? Hence calculate the length of the hypotenuse.

Ex. **758**. Find the hypotenuse of a right-angled triangle when the sides containing the right angle are

(i)   5 cm., 12 cm.,

(ii)   4·5 in., 6 in.,

(iii)   7·8 cm., 9·4 cm.,

(iv)   2·34 in., 4·65 in.,

(v)   4¼ miles, 5¾ miles,

(vi)   65 mm., 83·5 mm.

Ex. **759**. Find the remaining side and the area of a right-angled triangle, given the hypotenuse and one side, as follows:

(i)   hyp. = 15 cm.,   side = 12 cm.,

(ii)   hyp. = 6 in.,   side = 4 in.,

(iii)   hyp. = 8 in.,   side = 4 in.,

(iv)   hyp. = 160 mm.,   side = 100 mm.,

(v)   hyp. = 143 mm.,   side = 71·5 mm.

Ex. **760**. A flag-staff 40 ft. high is held up by several 50 ft. ropes; each rope is fastened at one end to the top of the flag-staff, and at the other end to a peg in the ground. Find the distance between the peg and the foot of the flag-staff.

Ex. **761**. Find the diagonal of a rectangle whose sides are 4 in. and 6 in.

Ex. **762**. Find the remaining side and the area of a rectangle, given diagonal = 10 cm., one side = 7 cm.

Ex. **763**. Find the diagonal of a square whose side is (i) 1 in., (ii) 5 in.

Ex. **764**. Find the side and area of a square whose diagonal is (i) 2 in., (ii) 14·14 cm.

Ex. **765**. Find the side of a rhombus whose diagonals are 16 cm., 12 cm.

Ex. **766**. Find the altitude of an isosceles triangle, given (i) base = 4 in., side = 5 in., (ii) base = 64 mm., side = 40 mm.

Ex. **767**. Find the altitude of an equilateral triangle of side 10 cm.

Ex. 768. In fig. 143, ABCD represents a square of side 3 in.; AE=AH=CF=CG=1 in. Prove that EFGH is a rectangle; find its perimeter and diagonal.

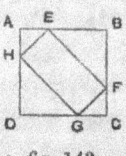

fig. 143.

Ex. 769. Find how far a traveller is from his starting point after the following journeys:—(i) first 10 miles N., then 8 miles E., (ii) first 8 miles E., then 10 miles N., (iii) 43 km. S.W. and 32 km. S.E., (iv) 14 miles S., 10 miles E., 4 miles N. (try to complete a right-angled triangle having the required line for hypotenuse), (v) 4 miles E., 6 miles N., 3 miles E., 1 mile N.

Ex. 770. (Inch paper.) If the coordinates of a point P are (1, 1) and of Q (2, 3), find the distance PQ. (PQ is the diagonal of a certain rectangle.)

Ex. 771. Given two squares of different sizes, show how to construct a square equal to (i) their sum, (ii) their difference.

Ex. 772. (Inch paper.) In each of the following cases find the distance between the pair of points whose coordinates are given:—(i) (2, 1) and (1, 3); (ii) (0, 0) and (3, 1); (iii) (2, 0) and (0, 3); (iv) (−1, −1) and (2, 1); (v) (−2, 2) and (1, −2).

Ex. 773. Newhaven is 90 miles N. of Havre, and 50 miles E. of Portsmouth. How far is it from Portsmouth to Havre?

Ex. 774. St Albans is 32 miles N. of Leatherhead, and Leatherhead is 52 miles from Oxford. Oxford is due W. of St Albans; how far is Oxford from St Albans?

Ex. 775. A ship's head is pointed N., and it is steaming at 15 miles per hour. At the same time it is being carried E. by a current at the rate of 4 miles per hour. How far does it actually go in an hour, and in what direction?

Ex. 776. A ladder 60 ft. long is placed against a wall with its foot 20 ft. from the foot of the wall. How high will the top of the ladder be?

Ex. 777. Find the distance between the summits of two columns, 60 and 40 ft. high respectively, and 30 ft. apart.

Ex. 778. What is the hypotenuse of a right-angled triangle whose sides are a and b in.?

Ex. **779**. What is the remaining side of a right-angled triangle which has hypotenuse$=x$ in. and one side $=y$ in.?

Ex. **780**. The edges of a certain cuboid (rectangular block) are 3″, 4″, 6″; find the diagonals of the faces.

Ex. **781**. A room is 18 ft. long, 14 ft. wide, 10 ft. high. Find the diagonals of the walls. Find the diagonal of the floor.

Ex. **782**. Find the length of a string stretched across the room in the preceding exercise, from one corner of the floor to the opposite corner of the ceiling.

Ex. **783**. Find the diagonal of the face of a cubic decimetre. Also find the diagonal of the cube.

Ex. **784**. Find the slant side of a cone of (i) height 5″, base-radius 3″; (ii) height 4·6 cm., base-radius 7·5 cm.; (iii) height 55 mm., base-diameter 46 mm.

Ex. **785**. Find the height of a cone of (i) slant side 10″, base-radius 4″; (ii) slant side 5·8 m., base-diameter 11 m.

Ex. **786**. Find the base-radius of a cone of (i) slant side 7 ft., height 5 ft.; (ii) slant side 11·3 cm., height 57 millimetres.

If further practice is needed, the reader may solve, by calculation, Exs. 416–422, 428.

†Ex. **787**. Prove I. 15 by means of Pythagoras' theorem. (See p. 77.)

†Ex. **788**. AD is the altitude of a triangle ABC. Prove that
$$AB^2 - AC^2 = BD^2 - CD^2.$$

Ex. **789**. In Ex. 788 let $AB=3$ in., $AC=2$ in., $BC=3$ in. Calculate $BD^2 - CD^2$. Hence find $BD - CD$.

$$[BD^2 - CD^2 = (BD - CD)(BD + CD) = (BD - CD)BC.]$$

Knowing $BD - CD$ and $BD + CD$, you may now find BD and CD. Hence find AD. Hence find area of △ABC.

†Ex. **790**. PQR is a triangle, right-angled at Q. On QR a point S is taken. Prove that $PS^2 + QR^2 = PR^2 + QS^2$.

†Ex. **791**. The diagonals of a quadrilateral ABCD intersect at right angles. Show that $AB^2 + CD^2 = BC^2 + DA^2$.

THEOREM 6.*

[CONVERSE OF PYTHAGORAS' THEOREM.]

If a triangle is such that the square on one side is equal to the sum of the squares on the other two sides, then the angle contained by these two sides is a right angle.

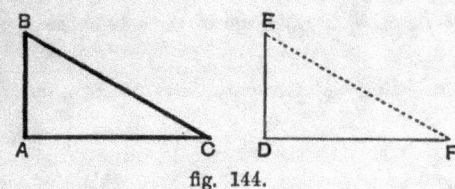

fig. 144.

Data   The triangle ABC is such that $BC^2 = AB^2 + AC^2$.

To prove that $\angle$ BAC is a right angle.

Construction   Construct a $\triangle$ DEF, to have DE = AB, DF = AC, and $\angle$ EDF a rt. $\angle$.

Proof   Since $\angle$ EDF is a right angle,   *Constr.*

$\therefore$ $EF^2 = DE^2 + DF^2$

$= AB^2 + AC^2$ *Constr.*

$= BC^2$,   *Data*

$\therefore$ EF = BC.

Hence, in the $\triangle$s ABC, DEF,

$\begin{cases} AB = DE, & \textit{Constr.} \\ AC = DF, & \textit{Constr.} \\ BC = EF, & \textit{Proved} \end{cases}$

$\therefore$ the triangles are congruent,

$\therefore$ $\angle$ BAC = $\angle$ EDF.

Now $\angle$ EDF is a right angle,   *Constr.*

$\therefore$ $\angle$ BAC is a right angle.

Q. E. D.

* This theorem does not appear in the schedules of the Cambridge Previous Examination and Oxford Responsions.

Ex. **792**.   Are the triangles right-angled whose sides are

(i) 8, 17, 15;   (ii) 12, 36, 34;   (iii) 25·5, 25·7, 3·2;

(iv) $4n$, $4n^2-1$, $4n^2+1$;   (v) m^2+n^2, m^2-n^2, $2mn$;   (vi) $a, b, a+b$?

Ex. **793**.   Bristol is 71 miles due W. of Reading; Reading is 55 miles from Northampton; Northampton is 92 miles from Bristol. Ascertain whether Northampton is due N. of Reading.

PROJECTIONS.

DEF.   If from the extremities of a line AB perpendiculars AM, BN are drawn to a straight line CD, then MN is called the **projection** of AB upon CD (figs. 145, 146).

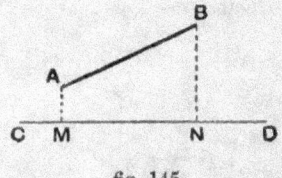   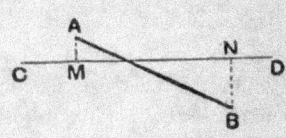

    fig. 145.   fig. 146.

¶Ex. **794**.   In fig. 140 name the projection of AB upon DE; of AE upon BC; of AC upon AL.

¶Ex. **795**.   In fig. 150 name the projection of AC upon BN; of BC upon NC.

¶Ex. **796**.   (On squared paper.)   What is the length of the projection (i) upon the axis of x, (ii) upon the axis of y, of the straight lines whose extremities are the points

(a) (2, 3) and (6, 6)?   (d) (−1, −3) and (3, 0)?

(b) (2, 4) and (6, 7)?   (e) (1, 1) and (5, 1)?

(c) (0, 0) and (4, 3)?   (f) (0, −2) and (0, 2)?

¶Ex. **797**.   In what case is the projection of a line equal to the line itself?

¶Ex. **798**.   In what case is the projection of a line zero?

¶Ex. **799**.   How does the projection of a line of given length alter as the slope of the line becomes more and more steep? Draw a graph showing how the projection varies as the angle of slope increases from 0° to 90°.

†Ex. **800.** **Prove that the projections on the same straight line of equal and parallel straight lines are equal.** (See fig. 147.)

†Ex. **801.** O is the mid-point of AB; the projections of A, B, O upon any line are P, Q, T. Prove that PT = QT.

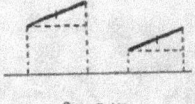

Ex. **802.** Prove that, if the slope of a line is 60°, its projection is half the line.

[Consider an equilateral triangle.]

fig. 147.

Ex. **803.** A pedestrian first ascends at an angle of 12° for 2000 yards and then descends at an angle of 9° for 1000 yards. How much higher is he than when he started? What horizontal distance has he travelled (i.e. what is the projection of his journey on the horizontal)?

Ex. **804.** The projections of a line of length l upon two lines at right angles are x, y. Prove that $x^2 + y^2 = l^2$.

NOTE. It may be necessary to produce the line upon which we project, e.g. if required to project AB upon CD in fig. 148, we must produce CD.

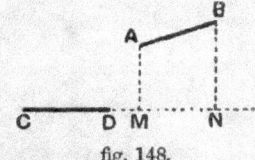

fig. 148.

EXTENSION OF PYTHAGORAS' THEOREM.

Suppose that two rods AB, AC are hinged together at A, and that their extremities B, C are connected by elastic.

As the angle BAC is increased from an acute angle to a right angle, and from a right angle to an obtuse angle, the elastic BC is stretched more and more. Thus, in fig. 149,

$$BC_1 < BC \text{ and } BC_2 > BC.$$

Now $BC^2 = CA^2 + AB^2$,

∴ $BC_1^2 = C_1A^2 + AB^2 -$ some area,

and $BC_2^2 = C_2A^2 + AB^2 +$ some area.

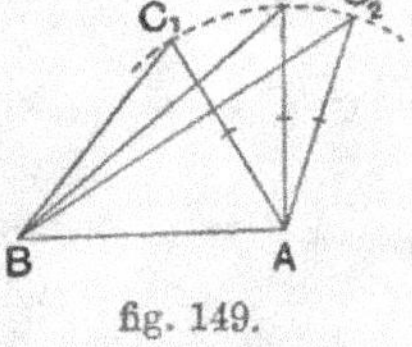

fig. 149.

The precise value of the quantity referred to as "some area" is given in the two following theorems.

Theorem 7.

In an obtuse-angled triangle, the square on the side opposite to the *obtuse* angle is equal to the sum of the squares on the sides containing the obtuse angle *plus* twice the rectangle contained by one of those sides and the projection on it of the other.

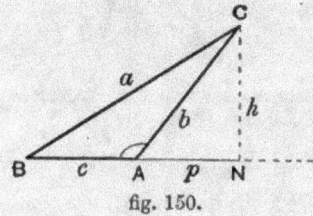

fig. 150.

Data   The $\triangle$ ABC has $\angle$ BAC obtuse.

CN is the perpendicular from C upon BA (produced),

∴ AN is the projection of AC upon BA.

Let BC=a units, CA=b units, AB=c units, AN=p units, CN=h units.

To prove that   BC^2 = CA^2 + AB^2 + 2AB . AN,

i.e. that $a^2 = b^2 + c^2 + 2cp$.

Proof   Since $\triangle$ BNC is right-angled,

∴ BC^2 = BN^2 + NC^2,   *Pythagoras*

i.e. $a^2 = (c + p)^2 + h^2$

$= c^2 + 2cp + p^2 + h^2$.

But $\triangle$ ANC is right-angled,

∴ $p^2 + h^2 = b^2$,   *Pythagoras*

∴ $a^2 = c^2 + 2cp + b^2$,

i.e. BC^2 = AB^2 + 2AB . AN + AC^2.

Q. E. D.

Theorem 8.

In any triangle, the square on the side opposite to an *acute* angle is equal to the sum of the squares on the sides containing that acute angle *minus* twice the rectangle contained by one of those sides and the projection on it of the other.

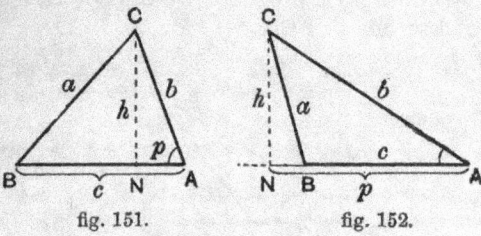

fig. 151.   fig. 152.

Data   The $\triangle$ ABC has $\angle$ BAC acute.

CN is the perpendicular from C upon AB (or AB produced),

$\therefore$ AN is the projection of AC upon AB.

Let BC $= a$ units, CA $= b$ units, AB $= c$ units, AN $= p$ units, CN $= h$ units.

To prove that   BC^2 = CA^2 + AB^2 − 2AB . AN,

i.e. that $a^2 = b^2 + c^2 - 2cp$.

Proof   Since $\triangle$ BNC is right-angled,

$\therefore$ BC^2 = BN^2 + NC^2,   *Pythagoras*

i.e. in fig. 151, $a^2 = (c - p)^2 + h^2$,

in fig. 152, $a^2 = (p - c)^2 + h^2$,

$\therefore$ in both figures,

$a^2 = c^2 - 2cp + p^2 + h^2$.

But $\triangle$ ANC is right-angled,

$\therefore p^2 + h^2 = b^2$,   *Pythagoras*

$\therefore a^2 = c^2 - 2cp + b^2$,

i.e. BC^2 = AB^2 − 2AB . AN + AC^2.

Q. E. D.

Ex. **805.**   Calculate BC when

(i)   AB=10 cm.,   AC=8 cm.,   ∠A=60°.   (See Ex. 802.)

(ii) AB=10 cm.,   AC=8 cm.,   ∠A=120°.

Ex. **806.**   Calculate the area of the rectangle referred to in the enunciation of II. 7, 8 for the following cases :

(i) $c=3$ in., $b=2$ in., $a=4$ in.

(ii) $c=3$ in., $b=2$ in., $a=2$ in.

Ex. **807.**   By comparing the square on one side with the sum of the squares on the two other sides, determine whether triangles having the following sides are acute-, obtuse-, or right-angled :

(i) 3, 4, 6;   (ii) 3, 4, 3;   (iii) 2, 3, 5;   (iv) 2, 3, 4;   (v) 12, 13, 6.

Ex. **808.**   Given four sticks of lengths 2, 3, 4, 5 feet, how many triangles can be made by using three sticks at a time? Find out whether each triangle is acute-, obtuse-, or right-angled.

¶ Ex. **809.**   What does II. 8 become if ∠B is a right angle?

¶ Ex. **810.**   Suppose that ∠A in fig. 150 becomes larger and larger till BAC is a straight line. What does II. 7 become in this case?

¶ Ex. **811.**   Suppose that ∠A in fig. 151 becomes smaller and smaller till C is on BA. What does II. 8 become in this case?

†Ex. **812.**   In the trapezium ABCD (fig. 153), prove that $AC^2+BD^2=AD^2+BC^2+2AB.CD$.
   [Apply II. 8 to Δ^s ACD and BCD.]

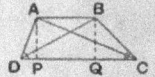

fig. 153.

†Ex. **813.**   D is a point on the base BC of an isosceles Δ ABC. Prove that $AB^2=AD^2+BD.CD$.

   (Let O be mid-point of BC, and suppose that D lies between B and O. Then $BD=BO-OD, CD=CO+OD=BO+OD$.)

†Ex. **814.**   ABC is an isosceles Δ (AB=AC); BN is an altitude. Prove that $2AC.CN=BC^2$.

†Ex. **815.**   BE, CF are altitudes of an acute-angled Δ ABC. Prove that $AE.AC=AF.AB$. [Write down two different expressions for BC^2.]

†Ex. **816.**   In the figure of Ex. 815, $BC^2=AB.FB+AC.EC$.

†Ex. **817.** **The sum of the squares on the two sides of a triangle ABC is equal to twice the sum of the squares on the median AD, and half the base. (Apollonius' theorem.)**
[Draw AN ⊥ to BC; apply ɪɪ. 7, 8 to △ˢ ABD, ACD.]

Ex. **818.** Use Apollonius' theorem to calculate the lengths of the three medians in a triangle whose sides are 4, 6, 7.

Ex. **819.** Repeat Ex. 818, with sides 4, 5, 7.

Ex. **820.** Calculate the base of a triangle whose sides are 8 cm. and 16 cm., and whose median is 12 cm. Verify graphically.

¶ Ex. **821.** What does Apollonius' theorem become if the vertex moves down (i) on to the base, (ii) on to the base produced?

ILLUSTRATION OF ALGEBRAICAL IDENTITIES BY MEANS OF GEOMETRICAL FIGURES.

The area of a rectangle a inches long and b inches broad is ab sq. inches.

¶ Ex. **822.** What is the area of a square whose side is x inches?

¶ Ex. **823.** Find an expression for the area of each of the following rectangles (do not remove the brackets):

(i) $(a+b)$ inches long, k inches broad;

(ii) $(a+b)$ cm. long, $(c+d)$ cm. broad;

(iii) $(a+b)$ cm. long, $(a-b)$ cm. broad.

¶ Ex. **824.** What is the area of a square whose side is $(a+b)$ inches? Is the answer equal to (a^2+b^2) sq. inches?

¶ Ex. **825.** Simplify the following expressions by removing brackets:

(i) $(a+b)(c+d)$,   (ii) $(a+b)^2$,   (iii) $(a-b)^2$.

It will now be shown how certain algebraic identities can be illustrated by geometrical figures.

(A)   Geometrical illustration of the identity

$$(a + b)\,k \equiv ak + bk.$$

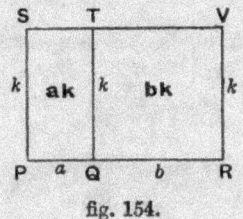

fig. 154.

The area of the whole figure is $(a + b)\,k$ units of area.

The areas of its parts are ak, bk units of area.

$$\therefore\ (a + b)\,k \equiv ak + bk.$$

Ex. 826. Give a geometrical illustration of the identity

$$(a+b+c)\ k \equiv ak+bk+ck$$

(i.e. draw a figure and give an explanation).

(B)   Geometrical illustration of the identity

$$(a + b)\,(c + d) \equiv ac + bc + ad + bd.$$

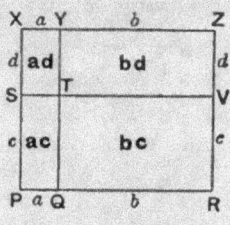

fig. 155.

The area of the whole figure is $(a + b)\,(c + d)$ units of area.

The areas of its parts are ac, bc, ad, bd units of area.

$$\therefore\ (a + b)\,(c + d) \equiv ac + bc + ad + bd.$$

(C) **Geometrical illustration of the identity**

$$(a+b)^2 \equiv a^2 + b^2 + 2ab.$$

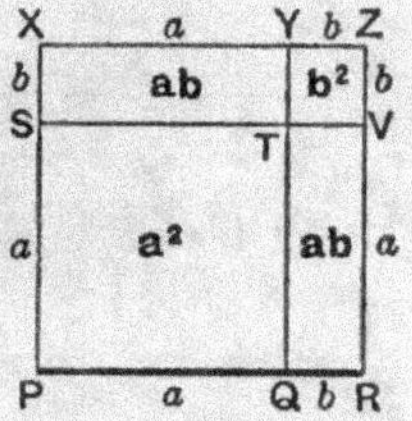

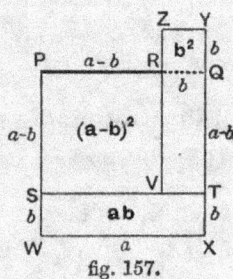

fig. 156.

The area of the whole figure is $(a+b)^2$ units of area.

The areas of its parts are a^2, b^2, ab, ab units of area.

$$\therefore \ (a+b)^2 \equiv a^2 + b^2 + 2ab.$$

(D) **Geometrical illustration of the identity**

$$(a-b)^2 \equiv a^2 + b^2 - 2ab.$$

fig. 157.

Let PQ $= a$ units of length.

From PQ cut off the length QR, containing b units.

Draw a figure as shown.

Then the whole figure contains $(a^2 + b^2)$ units of area.

In rect. VY, side YZ $= b$ units of length,

and side YT $= b + (a-b) = a$ units of length.

$\therefore$ Rect. VY contains ab units of area.

Now sq. SR $=$ whole fig. $-$ rect. WT $-$ rect. VY,

$$\therefore \ (a-b)^2 \equiv (a^2 + b^2) - ab - ab \equiv a^2 + b^2 - 2ab.$$

(E)   Geometrical illustration of the identity

$$a^2 - b^2 \equiv (a+b)(a-b).$$

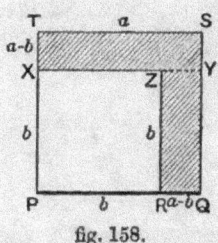

fig. 158.

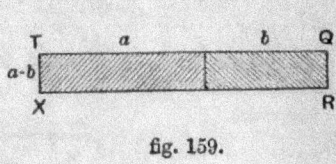

fig. 159.

Let PQ $= a$ units of length.

From PQ cut off the length PR, containing b units of length.

Draw fig. 158.

The area of the shaded part is $(a^2 - b^2)$ units of area.

Now this part is composed of the rectangles XS and RY.

The breadth of each is $(a-b)$ units of length; they might therefore be placed end to end, so as to form a single rectangle (as shown in fig. 159).

This single rectangle contains $(a+b)(a-b)$ units of area,

$$\therefore\ a^2 - b^2 \equiv (a+b)(a-b).$$

Ex. 827.   Prove algebraically that

$$(a+b+c)^2 \equiv a^2 + b^2 + c^2 + 2bc + 2ca + 2ab\,;$$

also give a geometrical illustration of the identity.

Ex. 828.   Illustrate the identity $(2x)^2 \equiv 4x^2$.

MISCELLANEOUS EXERCISES.

†Ex. **829**. A four-sided field is to be divided into two parts of equal area; prove the accuracy of the following construction. Draw a quadrilateral ABCD to represent the field; draw the diagonal AC; find E, the mid-point of AC; join BE, DE; then the areas ABED and CBED are equal.

†Ex. **830**. The area of a parallelogram of angle 30° is half the area of a rectangle with the same sides.

†Ex. **831**. O is any point on the diagonal BD of a parallelogram ABCD. EOF, GOH are parallel to AB, BC respectively. Prove that parallelogram AO = parallelogram CO.

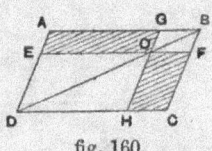

fig. 160.

†Ex. **832**. Any straight line drawn through the centre of a parallelogram (i.e. through the intersection of the diagonals) bisects the parallelogram.

Ex. **833**. Show how to bisect a parallelogram by a straight line drawn perpendicular to a side.

Ex. **834**. Show how to bisect a parallelogram by a straight line drawn through a given point.

†Ex. **835**. E is any point on the diagonal AC of a parallelogram ABCD. Prove that △ ABE = △ ADE.

†Ex. **836**. Produce the median BD of a triangle ABC to E, making DE = DB. Prove that △ EBC = △ ABC.

†Ex. **837**. P, Q are the mid-points of the sides BC, AD of the trapezium ABCD; EPF, GQH are drawn perpendicular to the base. Prove that trapezium = rectangle GF. (See fig. 161.)

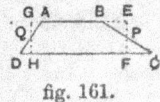

fig. 161.

†Ex. **838**. L, M are the mid-points of the parallel sides AB, CD of a trapezium ABCD. Prove that LM bisects the trapezium.

†Ex. **839**. In Ex. 888 O is the mid-point of LM; prove that any line through O which cuts AB, CD (not produced) bisects the trapezium.

†Ex. **840**. Prove that the area of the parallelogram formed by joining the mid-points of the sides of *any* quadrilateral ABCD (see Ex. 558) is half the area of the quadrilateral.

†Ex. **841.** The medians BD, CE of △ ABC intersect at G; prove that quadrilateral ADGE = △ BGC. (Add to each a certain triangle.)

†Ex. **842.** O is a point inside a triangle ABC such that the triangles BOC, COA, AOB are of equal area. Prove that O is the point of intersection of the medians of the triangle ABC.

†Ex. **843.** Draw a quadrilateral ABCD; from B and C draw BE and CF perpendicular to AD. Prove that the area of the quadrilateral ABCD is equal to the sum of the areas of the triangles ABF and ECD.
[First consider the case in which angles A and D are both acute.]

Ex. **844.** Perform, and prove, the following construction for erecting a perpendicular to a given straight line AB at its extremity A. Along AB mark off AC = 3 units. On AC as base construct a triangle ACD, having AD = 4, CD = 5. Then AD is perpendicular to AB. (Ancient Egyptian method.)

†Ex. **845.** ABC is a triangle, right-angled at A. On AB, AC respectively points X, Y are taken. Prove that $BY^2 + CX^2 = XY^2 + BC^2$.

†Ex. **846.** O is a point inside a rectangle ABCD. Prove that
$$OA^2 + OC^2 = OB^2 + OD^2.$$

†Ex. **847.** The base BC of an isosceles △ ABC is produced to D, so that CD = BC; prove that $AD^2 = AC^2 + 2BC^2$.

†Ex. **848.** A side PR of an isosceles △ PQR is produced to S so that RS = PR: prove that $QS^2 = 2QR^2 + PR^2$.

†Ex. **849.** A point moves so that the sum of the squares of its distances from two fixed points A, B remains constant; prove that its locus is a circle, having for centre the mid-point of AB.

Ex. **850.** Prove that the square on the difference of the sides of a right-angled triangle, together with twice the rectangle contained by the sides, is equal to the square on the hypotenuse. (Use Algebra.)

†Ex. **851.** Prove that, if in fig. 140 AD, FC intersect at O, BO bisects ∠ DOF.

†Ex. **852.** Prove that, if the sum of the squares on two opposite sides of a quadrilateral is equal to the sum of the squares on the two remaining sides, the diagonals of the quadrilateral must be at right angles.

†Ex. **853.** Find the condition that must exist in order that it may be possible to fold the four corners of a quadrilateral piece of paper flat on the paper so that the angular points meet in a point, and the paper is everywhere doubled.

†Ex. **854.** Any point D is taken on the base BC of an equilateral triangle ABC; prove that the sum of the rectangle contained by BD and DC and the square on AD is the same for all positions of D between B and C.

†Ex. **855.** A point P is taken within a triangle ABC such that when perpendiculars PM, PN are let fall on AB, AC respectively the rectangles PN . AC and PM . AB are equal. Prove that P lies on a fixed straight line.

†Ex. **856.** On the sides of a parallelogram ABCD are taken four points E, F, G, H at the same distance from the four corners in order: show that the area of the parallelogram EFGH is equal to the sum of the areas of the parallelograms AK, and CK, where K is the point in which parallels through E and F to the sides intersect.

Ex. **857.** Show how to describe a triangle whose angular points lie on three given parallel straight lines having given the area and the length of the straight line joining one of, the angular points to the middle point of the opposite side.

†Ex. **858.** Prove that, if O be any point in the plane of a parallelogram ABCD and the parallelograms OAEB, OBFC, OCGD, ODHA be completed, then EFGH will be a parallelogram whose area is double that of the parallelogram ABCD.

†Ex. **859.** Any point O is taken in the plane of a parallelogram ABCD: prove that the difference of the sum of the squares on OA and OC and the sum of those on OB and OD is half the difference of the squares on the diagonals AC and BD.

†Ex. **860.** A point O is taken within the parallelogram ABCD; prove that the difference of the areas of the triangles ODC and ODA is equal to the area of the triangle ODB.

†Ex. **861.** Two points P and Q are taken in the sides BC, CD respectively of the rectangle ABCD ; prove that the rectangle contained by BP and DQ, together with twice the triangle APQ, is equal to the rectangle ABCD.

†Ex. **862.** The points E and F are taken in the sides AB and AD respectively of the parallelogram ABCD ; EG drawn parallel to AD meets CD in G, and FH drawn parallel to AB meets BC in H. If K be any point in EF show that the triangles KBH and KDG are together equal in area to the triangle ECF.

10—2

Ex. **863**. Half the area of a square is cut away in the form of a square about a diagonal, one corner being common to the two squares. Prove that the remaining figure is divided by the diagonals of the original square into four parts which may be re-united into a square.

†Ex. **864**. If a quadrilateral is bisected as regards area by one diagonal, then the line joining the middle points of two opposite sides is also bisected by it.

†Ex. **865**. Prove that any quadrilateral is equal in area to a triangle whose sides are equal in length to the diagonals and double the line joining the middle points of two opposite sides.

†Ex. **866**. ABCD is a convex quadrilateral and AC, BD intersect at O. Show that the difference between the areas of AOD and BOC is equal to the area of a triangle of which AB, CD are sides and the angle between AB and CD is the included angle.

†Ex. **867**. The point D on the base BC of a triangle is such that the sum of the squares on AB and DC is equal to the sum of the squares on AC and DB. Prove that AD is perpendicular to BC.

†Ex. **868**. A point D inside a triangle ABC is such that the sum of the squares on DB and AC is equal to the sum of the squares on DC and AB. Prove that AD is at right angles to BC.

†Ex. **869**. If D be a point within a triangle ABC, and AD be produced to E so that DE is equal to DA, and BD be produced to F so that DF is equal to DB, show that the quadrilateral AFCE is equal in area to the triangle ABC.

†Ex. **870**. On the diagonal AB of a square two points P and Q are taken such that AQ, PB are each equal to a side of the square; prove that the square on PQ is twice that on AP.

†Ex. **871**. Prove that, if the perpendicular from the angle A on the side BC of the triangle ABC be greater than half the side BC, the angle A must be acute.

†Ex. **872**. A straight line DE is drawn to cut the base BC of a triangle ABC and is terminated by the sides AB, AC produced if necessary; prove that, if the difference of the areas of the triangles BCD, BCE is constant, the middle point of DE lies on one of two fixed straight lines.

BOOK III.

THE CIRCLE.

CONE.   SPHERE.   CYLINDER.

SECTION I. PRELIMINARY.

DEF. A **circle** is a line, lying in a plane, such that all points in the line are equidistant from a certain fixed point, called the **centre** of the circle.

In view of what has been said already about loci we may give the following alternative definition of a circle :

DEF. A **circle** is the locus of points in a plane that lie at a fixed distance from a fixed point (the centre). The fixed distance is called the **radius** of the circle.

The word "circle" has been defined above to mean a certain kind of curved line. The term is, however, often used to indicate the part of the plane inside this line. If any doubt exists as to the meaning, the line is called the **circumference** of the circle.

Two circles are said to be **equal** if they have equal radii.

Point and circle.   A point may be either outside a circle, on the circle or inside the circle.   The point will lie outside the circle if its distance from the centre > the radius; it will lie on the circle if its distance = the radius; it will lie inside the circle if the distance < the radius.

Straight line and circle.   A straight line cannot cut a circle in more than two points.   In fact, an unlimited straight line may

(i) cut a circle in two points, e.g. AB or CD in fig. 162. In this case the part of the line which lies inside the circle is called a **chord** of the circle.

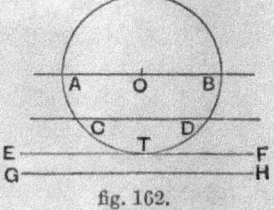

fig. 162.

(ii) The line may meet the circle in one point only; thus EF meets the circle in T. In this case the line is said to **touch** the circle; it is called a **tangent**; T is called the **point of contact** of the tangent.

The tangent lies entirely outside the circle and has one point, and one only, in common with the circle.   It is obvious that there is one and only one tangent which touches the circle at a given point.

(iii)   The line may lie entirely outside the circle, and have no point in common with the circle, e.g. GH in fig. 162.

A chord may be said to be the straight line joining two points on a circle.   If the chord passes through the centre it is called a **diameter**, e.g. AOB in fig. 162.

The length of a diameter is twice the length of the radius; all diameters are equal.

A chord divides the circumference into two parts called **arcs**. If the arcs are unequal, the less is called the **minor** arc and the greater the **major** arc.

Three letters are needed to name an arc completely; e.g. in fig. 162, CTD is a minor arc, CBD a major arc.

A diameter divides the circumference into two equal arcs, each of which is called a **semicircle**.

It will be proved below that the two semicircles are equal.

The term "semicircle" like the term "circle" is used in two different senses; sometimes in the sense of an arc (as in the definition); sometimes as the part of the plane bounded by a semi-circumference and the corresponding diameter.

A **segment** of a circle is the part of the plane bounded by an arc and its chord (fig. 163). A **sector** of a circle is the part of the plane bounded by two radii, and the arc which they intercept (fig. 163).

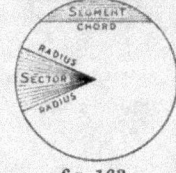

fig. 163.

¶Ex. **873.** A circular hoop is cut into two pieces; what is each called?

¶Ex. **874.** A penny is cut into two pieces by a straight cut; what is the shape of each piece?

¶Ex. **875.** What geometrical figure has the shape of an open fan?

¶Ex. **876.** A certain gun in a fort has a range of 5 miles, and can be pointed in any direction from 15° E. of N. to 15° W. of N. What is the shape of the area commanded by the gun?

SECTION II. CHORD AND CENTRE.

Symmetry of the circle. From what has been said about symmetry it will be seen that the circle is symmetrical about any diameter.

¶Ex. **877.** Draw a circle of about 3 in. radius; draw freehand a set of parallel chords (about 6); bisect each chord by eye. What is the locus of the mid-points of the chords?

¶Ex. **878.** Draw a circle and a diameter. This is an axis of symmetry. Mark four pairs of corresponding points. Is there any case in which a pair of corresponding points coincide? (*Freehand.*)

¶Ex. **879.** What axes of symmetry has (i) a sector, (ii) a segment, (iii) an arc, of a circle?

Theorem 1.

A straight line, drawn from the centre of a circle to bisect a chord which is not a diameter, is at right angles to the chord;

Conversely, the perpendicular to a chord from the centre bisects the chord.

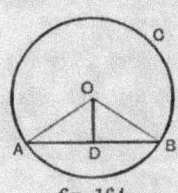

fig. 164.

(1) *Data*   OD is a straight line joining O, the centre of ⊙ABC, to D, the mid-point of the chord AB.

To prove that   OD is ⊥ to AB.

Construction   Join OA, OB.

Proof   In the △s OAD, OBD

$$\begin{cases} \text{OA} = \text{OB (radii),} \\ \text{OD is common,} \\ \text{AD} = \text{BD.} \end{cases}$$ *Data*

∴ the triangles are congruent,   I. 14.

∴ ∠ODA = ∠ODB,

∴ OD is ⊥ to AB.

(2) Converse Theorem.

Data   OD is a straight line drawn from O, the centre of ⊙ABC, to meet the chord AB at right angles in D.

To prove that   AD = BD.

Construction   Join OA, OB.

Proof   In the right-angled △s OAD, OBD

$\begin{cases} \angle s \text{ ODA, ODB are rt. } \angle s, & Data \\ OA = OB \text{ (radii)}, \\ OD \text{ is common}, \end{cases}$

∴ the triangles are congruent,   I. 15.

∴ AD = BD.   Q. E. D.

COR. A straight line drawn through the mid-point of a chord of a circle at right angles to the chord will, if produced, pass through the centre of the circle. (For only one perpendicular can be drawn to a given line at a given point in it.)

To find the centre of a given circle.

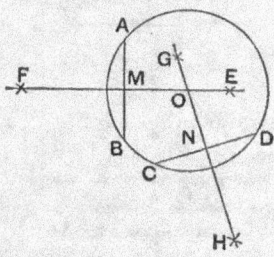

fig. 165.

Construction Draw any two chords AB, CD (not parallel).
          Draw EMF to bisect AB at right angles,
          and GNH to bisect CD at right angles.
          Let these straight lines meet at O.
          Then O is the centre of the circle.

Proof   Since EMF bisects chord AB at right angles,
          ∴ the centre must lie somewhere on EMF.   I. 25.
          Similarly the centre must lie somewhere on GNH.
          Hence the centre is at O, the point of intersection of EMF and GNH.

¶Ex. **880**. Why is it necessary that the chords AB, CD should not be parallel?

¶Ex. **881**. Describe five circles (in the same figure) to pass through two given points A, B, 6 cm. apart. (The centre must be equidistant from A and B; what is the locus of points equidistant from A and B?)

THEOREM 2.

There is one circle, and one only, which passes through three given points not in a straight line.

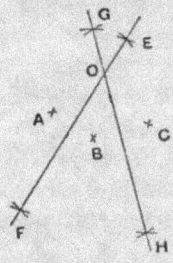

fig. 166.

Data   A, B, C are three points not in a straight line.

To prove that   one circle, and one only, can be drawn to pass through A, B and C.

Proof   It is only necessary to show that there is one point (and one only) equidistant from A, B, and C.

Now the locus of all points equidistant from A and B is FE, the perpendicular bisector of AB;   I. 25.
and the locus of all points equidistant from B and C is HG, the perpendicular bisector of BC.   I. 25.

These bisectors, not being parallel, will intersect.

Let the point of intersection be O.

The point O is equidistant from A and B ; also from B and C;

∴ O is equidistant from A, B and C ;

and there is no other point equidistant from A, B and C.

Hence a circle with centre O and radius OA will pass through A, B and C;

and there is no other circle passing through A, B and C.

                                                                        Q. E. D.

Cor. 1.   Two circles cannot intersect in more than two points.

For if the two circles have three points in common, they have the same centre and radius, and therefore coincide.

Cor. 2.   The perpendicular bisectors of AB, BC, and CA meet in a point.

¶Ex. **882.** How would the proof of III. 2 fail if A, B, C were in a straight line?

†Ex. **883.** Prove Cor. 2. (Let two of the bisectors meet at a point O; then prove that O lies on the third bisector.)

Ex. **884.** With a fine-pointed pencil trace round part of the edge of a penny, or better still the curved edge of your protractor, so as to obtain an arc of a circle. (Take care to keep the pencil perpendicular to the paper.) Complete the circle by finding the centre.

DEF. If a circle passes through all the vertices of a polygon, the circle is said to be **circumscribed** about the polygon; and the polygon is said to be **inscribed** in the circle (fig. 167).

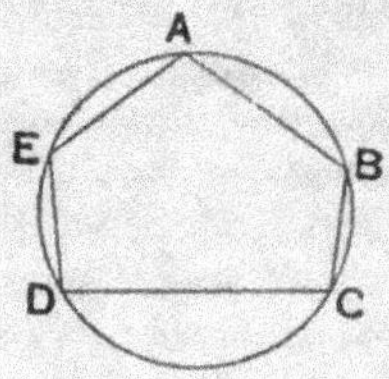

fig. 167.

DEF. If a circle touches all the sides of a polygon, the circle is said to be **inscribed** in the polygon; and the polygon is said to be **circumscribed** about the circle (fig. 168).

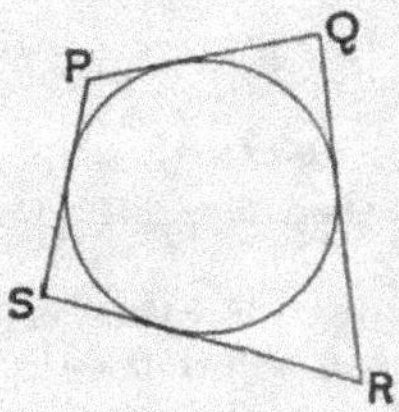

To circumscribe a circle about a given triangle.

This is the same problem as that of describing a circle to pass through three given points, namely the three vertices of the triangle (see III. 2).

fig. 168.

DEF. The centre of the circle circumscribed about a triangle is called the **circumcentre** of the triangle.

Notice that, though the perpendicular bisectors of all three sides pass through the circumcentre, yet it is not necessary to draw more than *two* of these bisectors in order to find the centre.

Ex. **885.** (Inch paper.) Draw a circle to pass through the points $(0, 3)$, $(2, 0)$, $(-1, 0)$, and measure its radius.

Ex. **886.** (Inch paper.) Draw the circumcircle of the triangle whose vertices are $(0, 2)$, $(4, 0)$, $(-1, 0)$, and find its radius.

Ex. **887.** (Inch paper.) Find the circumradius, and the coordinates of the circumcentre of $(0, 1)$, $(3, 0)$, $(-3, 0)$.

Ex. **888.** Mark four points (at random) on plain paper, and find out whether it is possible to draw a circle through all four.

¶Ex. **889**. Can a circle be circumscribed about a rectangle?

¶Ex. **890**. Draw a parallelogram (not rectangular) and try if a circle can be circumscribed about it.

†Ex. **891**. If a chord cuts two concentric circles in A, B ; C, D, then AC = BD. (Draw perpendicular from centre on to chord.)

†Ex. **892**. From a point O outside a circle two equal lines OP, OQ are drawn to the circumference. Prove that the bisector of ∠ POQ passes through the centre of the circle. (Join PQ.)

Ex. **893**. O is a point 4 inches from the centre of a circle of radius 2 inches. Show how to construct with O as vertex an isosceles triangle having for base a chord of the circle, and a vertical angle of 50°. How many solutions are there? (*Freehand.*)

†Ex. **894**. If a polygon is such that the perpendicular bisectors of all the sides meet in a point, a circle can be circumscribed round the polygon.

Section III*. Arcs, Angles, Chords.

Theorem 3.

In equal circles (or, in the same circle)

(1) **if two arcs subtend equal angles at the centres, they are equal.**

(2) *Conversely,* **if two arcs are equal, they subtend equal angles at the centres.**

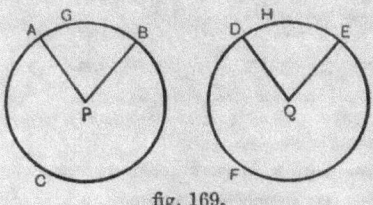

fig. 169.

(1) *Data*   ABC, DEF are equal ⊙s.

The arcs AGB, DHE subtend equal ∠ s APB, DQE at the centres P, Q.

---

* This section may be omitted at first reading, with the exception of Theorem 5 and the exercises which follow (pp. 163—166).

To prove that arc AGB = arc DHE.

Proof Apply ⊙DEF to ⊙ABC, so that centre Q may fall on centre P.

Since the ⊙s are equal, the circumference of ⊙DEF falls on the circumference of ⊙ABC.

Make ⊙DEF revolve about the centre till QD falls along PA.

Then, since ∠DQE = ∠APB, *Data*

QE falls along PB, and since the circumferences coincide, D coincides with A, and E with B.

∴ arc DHE coincides with arc AGB.

∴ arc DHE = arc AGB.

(2) CONVERSE THEOREM.

Data arc AGB = arc DHE.

To prove that ∠s APB, DQE, subtended by these arcs at the centres, are equal.

Proof Apply ⊙DEF to ⊙ABC, so that centre Q may fall on centre P.

Since the ⊙s are equal, the circumference of ⊙DEF falls on the circumference of ⊙ABC.

Make ⊙DEF revolve about the centre till D coincides with A.

Then, since arc DHE = arc AGB, *Data*

E coincides with B.

∴ QD coincides with PA, and QE with PB,

∴ ∠DQE = ∠APB. Q. E. D.

COR. Equal angles at the centre determine equal sectors.

†Ex. **895.** Show how to bisect a given arc of a circle. Give a proof.

†Ex. **896.** P, A, B are points on a circle whose centre is O; PA=PB. Prove that P is the mid-point of arc AB; and that OP bisects AB.

†Ex. **897.** PQ, PR are a chord and a diameter meeting at a point P in the circumference. Prove that the radius drawn parallel to PQ bisects the arc QR.

†Ex. **898.** P is a point on the circumference equidistant from the radii OA, OB. Prove that arc AP=arc BP.

Theorem 4.

In equal circles (or, in the same circle)

(1)   **if two chords are equal, they cut off equal arcs.**

(2)   *Conversely,* **if two arcs are equal, the chords of the arcs are equal.**

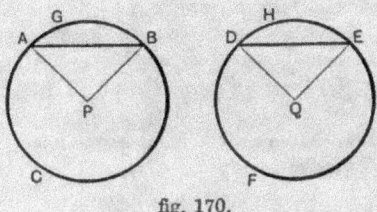

fig. 170.

(1) *Data*   ABC, DEF are equal ⊙s; their centres are P and Q.
                        Chord AB = chord DE.

To prove that   arc AGB = arc DHE, and arc ACB = arc DFE.

Construction   Join PA, PB; QD, QE.

Proof   In the △s APB, DQE

$\left\{\begin{array}{l} AB = DE, \\ AP = DQ \text{ (radii of equal ⊙s),} \\ BP = EQ \text{ (radii of equal ⊙s).} \end{array}\right.$ *Data*

∴ the triangles are congruent,   I. 14.
                ∴ ∠ APB = ∠ DQE,
            ∴ arc AGB = arc DHE.   III. 3.

Again, whole circumference of ⊙ ABC = whole circumference of ⊙ DEF.

∴ the remaining arc ACB = the remaining arc DFE.

(2) Converse Theorem.

Data   arc AGB = arc DHE.

To prove that   chord AB = chord DE.

Construction   Join PA, PB; QD, QE.

Proof   Since arc AGB = arc DHE,   *Data*

∴ ∠ APB = ∠ DQE,   III. 3.

∴ in the △s APB, DQE

$\begin{cases} AP = DQ, \\ BP = EQ, \\ ∠ APB = ∠ DQE. \end{cases}$

∴ the triangles are congruent,   I. 10.

∴ chord AB = chord DE.

Q. E. D.

†Ex. **899**. A quadrilateral ABCD is inscribed in a circle, and AB = CD. Prove that AC = BD.

†Ex. **900**. Prove the converse, in III. 4, by superposition. Also try to prove the direct theorem by superposition, and point out where such a proof fails.

Note on the case of "the same circle." Ths. 3 and 4 have been proved for equal circles. To see that they apply also to "the same circle," regard the circle as composed of two equal circles superimposed.

To place in a circle a chord of given length.

Adjust the compasses to the given length. With a point A on the circle as centre draw an arc cutting the circle in B. Then AB will be the chord required.

Ex. **901**. Place in a circle, end to end, 6 chords each equal to the radius.

Ex. **902**. Place in a circle, end to end, 12 chords each equal to ½ the radius.

Ex. **903**. Draw a circle of radius 5 cm. Place in the circle a number of chords of length 8 cm. Plot the locus of their middle points.

Ex. **904**. Show how to construct an isosceles triangle, given the base and the radius of the circumscribing circle. (Which will you draw first—the base, or the circle?)

†Ex. **905**. In a circle are placed, end to end, equal chords PQ, QR, RS, ST. Prove that PR = QS = RT.

To inscribe a regular hexagon in a circle.

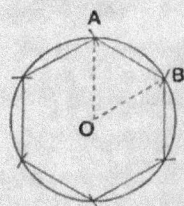

fig. 171.

In the circle place a chord AB, equal to the radius.

Join A, B to O, the centre.

Then △OAB is equilateral,

∴ ∠AOB = 60°.

Place end to end in the circle 6 chords each equal to the radius.

Each chord subtends 60° at the centre,

∴ the total angle subtended by the 6 chords is 360°.

In other words, the 6 chords form a closed hexagon inscribed in the circle.

Since each side of the hexagon = the radius, the hexagon is equilateral; and since each angle of the hexagon = 120°, the hexagon is equiangular,

∴ the hexagon is regular.

†Ex. **906.** The side of an isosceles triangle of vertical angle 120° is equal to the radius of the circumcircle.

CIRCUMFERENCE OF CIRCLE.

You will have found in the laboratory that the circumference of a circle contains the diameter about 3·14 times. The ratio of circumference to diameter has been worked out to 700 places of decimals and begins thus

3·1415926535...

For the sake of brevity this number is denoted by the Greek letter π; a useful approximation is $\frac{2\,2}{7}$ or $3\frac{1}{7}$.

Ex. **907**. Without assuming the value of π, prove that the circumference of a circle is > three times the diameter, by inscribing a hexagon in the circle.

Ex. **908**. Without assuming the value of π, prove that the circumference is < four times the diameter by circumscribing a square round the circle.

The ratios of the perimeters (or circumferences) of regular polygons to the diameters of their circumscribing circles are shown in the following table:

Table showing the perimeters of regular polygons inscribed in a circle of radius 5 cm.

No. of sides	Perimeter in centimetres	Ratio of perimeter to diameter
3	25·98	2·598
4	28·29	2·829
5	29·39	2·939
6	30·00	3·000
7	30·38	3·038
8	30·61	3·061
9	30·78	3·078
10	30·90	3·090

It will be noticed that the ratio increases with the number of sides, being always less than π. If the number of sides is very great, the ratio is very nearly equal to π. E.g. for a polygon of 384 sides the ratio is 3·14156......

Ex. **909**. By how much per cent. does the perimeter of a regular decagon inscribed in a circle differ from the perimeter of the circle? Give the result to one significant figure.

We have seen that

circumference of circle $=$ diameter $\times \pi$

$$= \text{radius} \times 2\pi$$

$$= 2\pi r, \text{ where } r \text{ is the radius.}$$

Ex. **910**. Calculate the circumference of a circle whose radius is (i) 7 in., (ii) 35 miles.   (Take $\pi = \frac{22}{7}$.)

Ex. **911**. Calculate to three significant figures the circumference of a half-penny (diameter 1 inch).

Ex. **912**. Calculate to three significant figures the circumference of the earth, measured round the equator, taking radius = 3963 miles.

Ex. **913**. How far does a wheel roll in one revolution if its diameter is 28 in.?

Ex. **914**. In fig. 172, AD is divided into three equal parts and all the arcs are semicircles; show that the four curved lines which connect A with D are of equal length.

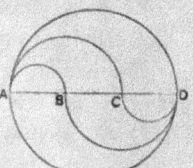

fig. 172.

Ex. **915**. Calculate the radius of a circle of circumference (i) 22 ft., (ii) 40 ft. (to three significant figures).

Ex. **916**. A bicycle wheel makes 7200 turns in an hour while the cyclist is riding 10 miles an hour: what is the diameter of the wheel (to the nearest inch)?

Def. If an arc of a circle subtends, say, 35° at the centre, it is called **an arc of 35°.**

Ex. **917**. What fractions of a circumference are arcs of 90°, 60°, 120°, 1°, 35°, 300°?

Ex. **918**. Calculate the length of an arc of 60° in a circle of radius 7 cm.   What is the length of the chord of this arc?   Find, to three significant figures, the ratio $\dfrac{arc}{chord}$: also the difference of arc and chord.

Ex. **919**. The circumference of a circle is 7·82 in. and the length of a certain arc is 1·25 in.   What decimal of the circumference is the arc? What angle does the arc subtend at the centre?

Ex. **920**. The radius of a circle is 10 cm.; a piece of string as long as the radius is laid along an arc of the circle; what angle does it subtend at the centre? Also find the angle subtended at the centre by a *chord* of 10 cm.

Ex. **921**. Find the length of the minor and major arcs cut off from a circle of radius 7 cm. by a chord of 7 cm.

Ex. **922**. Find the lengths of the two arcs cut from a circle of diameter 4·37 in. by a chord of 4 in.   (Measure the angle at the centre.)

Theorem 5.

In equal circles (or, in the same circle)

(1) **equal chords are equidistant from the centres.**

(2) *Conversely*, **chords that are equidistant from the centres are equal.**

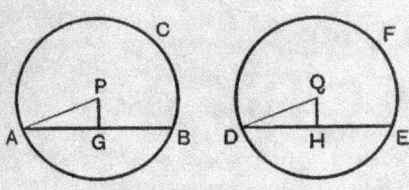

fig. 173.

(1) *Data*   ABC, DEF are equal circles; their centres are P and Q.
              Chord AB = chord DE.

          PG, QH are perpendiculars from the centres P, Q upon
          the chords AB, DE.

To prove that   PG = QH.

Construction   Join PA, QD.

Proof   Since PG is $\perp$ to AB,

                      $\therefore$ AG = BG,   III. 1.

                      $\therefore$ AG = $\frac{1}{2}$ AB.

                Sim^{ly} DH = $\frac{1}{2}$ DE.

                  But AB = DE,   *Data*

                      $\therefore$ AG = DH.

          In the right-angled $\triangle^s$ APG, DQH,

              $\Big\{$ $\angle^s$ G and H are rt. $\angle^s$,   *Constr.*

                    AP = DQ,   *Data*

                    AG = DH,   *Proved*

       $\therefore$ the triangles are congruent,   I. 15.

                    $\therefore$ PG = QH.

                              11—2

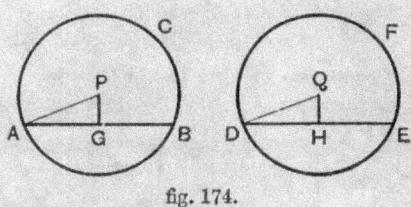

fig. 174.

(2) Converse Theorem.

Data   PG = QH.

To prove that   chord AB = chord DE.

Proof   In the right-angled △ˢ APG, DQH,

$$\begin{cases} ∠ˢ \text{ G and H are rt. } ∠ˢ, & \textit{Constr.} \\ AP = DQ, & \textit{Data} \\ PG = QH, & \textit{Data} \end{cases}$$

∴ the triangles are congruent,   I. 15.

∴ AG = DH.

But AB = 2AG,   DE = 2DH,

∴ AB = DE.

Q. E. D.

†Ex. **923.** Prove III. 5 by means of Pythagoras' theorem.

Ex. **924.** Calculate the distances from the centre of a circle (radius 5 cm.) of chords whose lengths are (i) 8 cm., (ii) 6 cm., (iii) 5 cm.

Ex. **925.** Calculate the lengths of chords of a circle (radius 2·5 in.) whose distances from the centre are (i) 2 in., (ii) 1·5 in., (iii) 1 in.

Ex. **926.** Find the locus of the mid-points of chords 6 cm. in length in a circle of radius 5 cm.

†Ex. **927.** Prove that the locus of the middle points of a set of equal chords of a circle is a concentric circle.

Ex. **928.** A chord CD of a circle, whose centre is O, is bisected at N by a diameter AB. OA=OB=5 cm., ON=4 cm. Calculate CD, CA, CB.

Ex. **929**. The lengths of two parallel chords of a circle of radius 6 cm. are 10 cm. and 6 cm. respectively. Calculate the distance between the chords. (There are two cases.)

Ex. **930**. Calculate the radius of a circle, given that a chord 3 in. long is 2 in. from the centre.

Ex. **931**. What is the radius of a circle when a chord of length $2l$ is at distance d from the centre?

Ex. **932**. Given that a chord 12 cm. long is distant 2·5 cm. from the centre, calculate (i) the length of a chord distant 5 cm. from centre, (ii) the distance from the centre of a chord 6 cm. long.

Ex. **933**. A wooden ball of 4″ radius is planed down till there is a flat circular face of radius 2″. If the block is now made to stand on the flat face, how high will it stand?

Ex. **934**. The distance from the centre of the earth of the plane of the Arctic circle is 3700 miles (to the nearest 100 miles); the radius of the earth is 4000 miles. Find the radius of the Arctic circle.

Ex. **935**. A ball of radius 4 cm. floats in water immersed to the depth of $\frac{1}{4}$ of its diameter. Calculate the circumference of the water-line circle.

Ex. **936**. The diameter of an orange is 4″, and the thickness of the rind is $\frac{1}{4}$″. A piece is sliced off just grazing the flesh; find the radius of the piece.

†Ex. **937**. If two chords make equal angles with the diameter through their point of intersection, they are equal.
[Prove that they are equidistant from the centre.]

†Ex. **938**. A straight line is drawn cutting two equal circles and parallel to the line joining their centres; prove that the chords intercepted by the two circles are equal.

†Ex. **939**. A straight line is drawn cutting two equal circles, and passing through the point midway between their centres. Prove that the chords intercepted by the two circles are equal.

Ex. **940**. Show how to draw a chord of a circle (i) equal and parallel to a given chord, (ii) equal and perpendicular to a given chord, (iii) equal to a given chord and parallel to a given line.

†Ex. **941**. If two chords are at unequal distances from the centre, the nearer chord is longer than the more remote.

†Ex. **942.**   State and prove the converse of Ex. 941.

†Ex. **943.**   The shortest chord that can be drawn through a point inside a circle is that which is perpendicular to the diameter through the point.

[Prove that it is furthest from the centre.]

Ex. **944.**   Calculate the length of (i) the longest, (ii) the shortest chord of a circle, radius r, through a point distant d from the centre.

SECTION IV.   THE TANGENT.

The meaning of the term tangent has been explained on p. 150. It may be defined as follows:

DEF.   A **tangent** to a circle is a straight line which, however far it may be produced, has one point, and one only, in common with the circle.

The tangent is said to **touch** the circle; the common point is called the **point of contact.**

We shall assume that at a given point on a circle there is one tangent and one only.

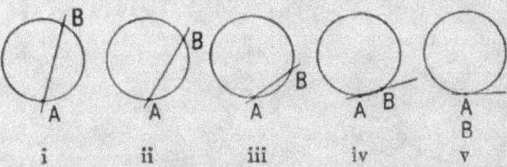

fig. 175.

The tangent may also be viewed in another way suggested by fig. 175. Figs. 175 (i—iv) show four positions of a chord AB (produced both ways).   Looking at the figures from left to right, the chord is seen to be turning about the point A; as it turns, the second point of intersection, B, comes nearer and nearer to A until in fig. v, B has coincided with A, and the chord has become the tangent at A.

A tangent therefore may be regarded as the **limit of a chord** whose two points of intersection with the circle have come to coincide.

Fig. 176 suggests another way in which the chord may approach its limiting position—the tangent.

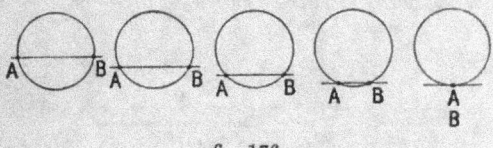

fig. 176.

¶ **Ex. 945.** In fig. 176, what becomes of the theorem that "the perpendicular from the centre on a chord bisects the chord" when B comes to coincide with A?

In fig. 177, i, ii, the triangle OAB is isosceles;

$$\therefore \angle OAB = \angle OBA, \quad \therefore \angle OAP = \angle OBQ.$$

When the chord attains its limiting position, the tangent, in fig. iii, we are left with $\angle OAP = \angle OAQ$. $\therefore$ the radius is perpendicular to the tangent at its extremity.

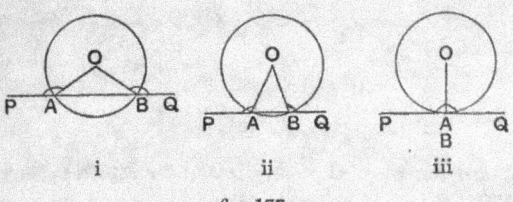

i   ii   iii

fig. 177.

We now give a proof of this theorem that does not involve the 'limit' idea.

THEOREM 6.

The tangent at any point of a circle and the radius through the point are perpendicular to one another.

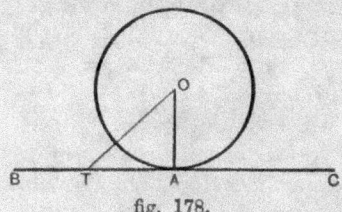

fig. 178.

Data   O is the centre of ⊙; A is a point on the circumference; BC is the tangent at A.

To prove that   BC and OA are ⊥ to one another.

Construction   If OA be not ⊥ to BC, draw OT ⊥ to BC.

Proof   Since ∠ OTA is a rt. ∠ ,   *Constr.*

∴ OT < OA,   (see p. 87)   I. 21.

∴ T is inside the circle,

∴ the tangent AT, if produced, will cut the circle in another point.

This is impossible,   *Def.*

∴ OA *is* ⊥ to BC,

∴ the tangent at A and the radius through A are ⊥ to one another.

Q. E. D.

Cor. A straight line drawn through the point of contact of a tangent at right angles to the tangent will, if produced, pass through the centre of the circle.

To draw the tangent to a circle at a given point on the circle.

Join the point to the centre, and draw a straight line through the point perpendicular to the radius.

The proper method of **drawing a tangent to a circle from an external point** cannot be explained at the present stage, as it depends on a proposition that has not yet been proved. In the meantime it will be sufficient to draw the tangent from an external point with the ruler (by eye). It is not possible to distinguish the point of contact accurately without further construction; to find this point, drop a perpendicular upon the tangent from the centre; the foot of this perpendicular is the point of contact.

This method is accurate enough for many purposes; the student is warned, however, that it would not be accepted in most examinations. The correct construction is given on page 198.

† Ex. **946. Prove that the two tangents drawn to a circle from a point A are (i) equal, (ii) equally inclined to AO.** (Fig. 179.)

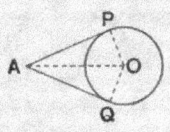

fig. 179.

Ex. **947.** P is 4 in. distant from O, the centre of a circle of radius 3 in. From P draw a tangent with your ruler. Determine T, the point of contact, (i) by eye, (ii) by drawing a perpendicular from O.

Calculate PT, the length of the tangent (using Pythagoras' theorem). Verify by measurement.

Ex. **948.** Calculate the lengths of the tangents to a circle of radius r from a point distant d from the centre when (i) $r=6$ cm., $d=8$ cm.; (ii) $r=1$ in., $d=5$ in.

Ex. **949.** At a point A of a circle (radius r, centre O) is drawn a tangent AP of length l; find OP.

Ex. **950.** At a point P on the circumference of a circle of radius 4 cm. is drawn a tangent PT 3 cm. in length. Find the locus of T as P moves round the circle.

Ex. **951.** Two circles, of radii 3 and 2 in., are concentric. Calculate the length of a chord of the outer circle which touches the inner.

Ex. **952.** Prove that all chords 8 cm. long of a circle of radius 5 cm. touch a certain concentric circle ; find its radius.

¶ Ex. **953.** If fig. 179 were spun about OA, what figure would be generated (i) by the circle, (ii) by AP, (iii) by PQ? Hence find the locus of the points of contact of tangents from a fixed point to a fixed sphere.

Ex. **954.** If P is a point 25 cm. from the centre of a sphere of radius 20 cm., what is the length of a tangent line from P to the sphere? Are all the tangent lines from P equal?

† Ex. **955.** All chords of a circle which touch an interior concentric circle are equal, and are bisected at the point of contact.

† Ex. **956.** PQRS **is a quadrilateral circumscribed about a circle. Prove that** PQ + RS = QR + SP. (See fig. 168.)

† Ex. **957.** Draw a circle and circumscribe a parallelogram about it. Prove that the parallelogram is necessarily a rhombus (use Ex. 956).

† Ex. **958.** Prove that the point of intersection of the diagonals of a rhombus is equidistant from the four sides.

¶ Ex. **959.** Draw a quadrilateral ABCD. What is the locus of the centres of circles touching AB, BC; touching BC, CD? Draw a circle to touch AB, BC and CD. Does it touch DA? What relation must hold between the sides of a quadrilateral in order that it may be possible to inscribe a circle in it?

† Ex. **960.** ABCDEF is an irregular hexagon circumscribed about a circle; prove that AB + CD + EF = BC + DE + FA.

† Ex. **961.** Two parallel tangents meet a third tangent at U, V; prove that UV subtends a right angle at the centre.

† Ex. **962.** The angles subtended at the centre of a circle by two opposite sides of a circumscribed quadrilateral are supplementary.

† Ex. **963.** A is a point outside a circle, of centre O. With centre O and radius OA describe a circle. Let OA cut the smaller circle in B. Draw BC perpendicular to OB, cutting the larger circle in P, Q. Let OP, OQ cut the smaller circle in S, T. Prove that AS, AT are tangents to the smaller circle. (This is Euclid's construction for tangents from an external point.)

†Ex. **964.** A chord makes equal angles with the tangents at its extremities.

Ex. **965.** Each of the tangents, TA, TB, at the ends of a certain chord AB is equal to the chord; find the angle between the tangents, and the angle subtended at the centre by the chord.

†Ex. **966.** In fig. 179, the angles PAQ, POQ are supplementary.

Ex. **967.** Show how to draw a tangent to a given circle (i) parallel to a given line, (ii) perpendicular to a given line, (iii) making a given angle with a given line.

Ex. **968.** Show how to draw two tangents to a circle inclined to one another at a given angle.

†Ex. **969.** The area of any polygon circumscribing a circle is equal to half the product of the radius of the circle, and the perimeter of the polygon. (Divide the polygon into triangles, with the centre for vertex.)

†Ex. **970.** AOB is a diameter of a circle whose centre is O. At C in the circumference a tangent CE is drawn. OE parallel to AC meets this tangent at E. Prove that EB touches the circle.

†Ex. **971.** A point P on a circle is joined to the extremity A of a diameter AB; the line AP, produced if necessary, cuts the diameter at right angles to AB in Q, and QC is a perpendicular on the tangent at B, meeting it in C. Prove that PC is the tangent at P.

†Ex. **972.** Through a fixed point A on a given circle a line is drawn cutting the circle in P, and AP is produced to Q so that PQ is of constant length. Show that the line through Q perpendicular to AQ touches a fixed circle of which the other end of the diameter through A is the centre.

†Ex. **973.** A straight line PQ is drawn through a point P on the circumference of a circle, making a constant angle with the tangent at P. Prove that, as P moves round the circumference, PQ touches a fixed circle.

¶Ex. **974.** What is the locus of the centres of circles touching two lines which cross at an angle of 60°? (Remember that two lines form four angles at a point.) Draw a number of such circles.

To inscribe a circle in a given triangle.

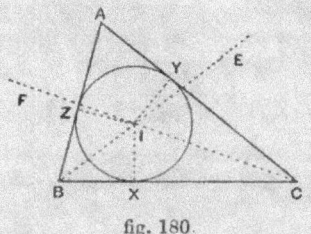

fig. 180.

Construction   It is necessary to find a point equidistant from the three straight lines AB, BC, CA.

Draw BE, CF to bisect the angles ABC, ACB respectively.

Let these lines intersect at I.

Then I is the centre of the inscribed circle.

Proof   Every point on BE is equidistant from AB and BC, and every point on CF is equidistant from BC, CA.   I. 26.

Therefore I is equidistant from AB, BC and CA.

From I draw IX, IY, IZ ⊥ to BC, CA, AB respectively.

Then IX = IY = IZ.

Therefore a circle described with I as centre and IX as radius will pass through X, Y, Z. Also BC, CA, AB will be tangents at X, Y, Z.   (Why?)

This circle is the **inscribed** circle of the triangle ABC.

Ex. **975**. Draw the inscribed circle of a triangle whose sides are (i) 5, 6, 7 in., (ii) 8, 6, 8 cm. Measure the radii of the circles.

†Ex. **976. The bisectors of the three angles of a triangle meet in a point.**

(Join IA, and prove that IA bisects ∠ A.)

The escribed circles of a triangle.

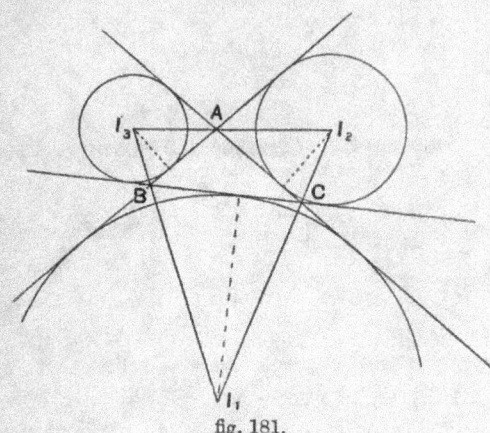

fig. 181.

Draw BI_1, CI_1 to bisect the angles exterior to ABC and BCA.

Then I_1 is equidistant from BC and AB, AC (produced).

Drop a perpendicular from I_1 to BC. A circle drawn with I_1 as centre and this perpendicular as radius will touch the side BC and the sides AB, AC produced. This circle is called an **escribed** circle of the triangle. There are three such circles (see fig. 181).

Ex. **977.** Draw the escribed circles of a triangle whose sides are 3, 4, 5 cm. Measure the radii.

†Ex. **978.** Prove that the internal bisector of ∠ A and the external bisectors of ∠ˢ B and C meet in a point.

†Ex. **979.** Prove that AII_1 is a straight line. (I is the centre of the inscribed circle.)

It has been shown that, in general, four circles can be drawn to touch three unlimited straight lines, namely the inscribed and escribed circles of the triangle which the three lines enclose.

¶Ex. **980.** How many circles can be drawn to touch two parallel straight lines and a third straight line cutting them?

¶Ex. **981.** How many circles can be drawn to touch three straight lines which intersect in a point?

¶Ex. **982.** How many circles can be drawn to touch three parallel straight lines?

SECTION V. CONTACT OF CIRCLES.

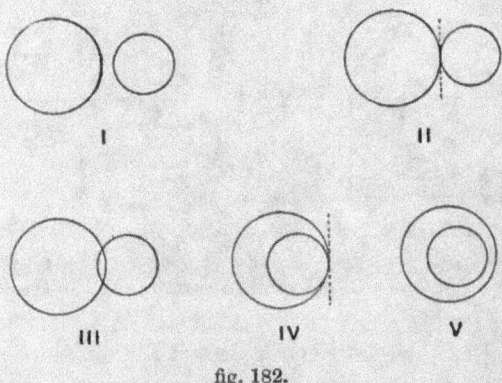

fig. 182.

The different relative positions which are possible for two circles are shown in fig. 182.

In Case II the circles are said to **touch externally**, in Case IV to **touch internally.** The formal definition of contact of circles is as follows:

DEF.   If two circles touch the same line at the same point, they are said to touch one another.

¶Ex. **983.** Draw a figure showing the different relative positions possible for two equal circles.

¶Ex. **984.** Describe in words each of the relative positions shown in fig. 182.

THEOREM 7.

If two circles touch, the point of contact lies in the straight line through the centres.

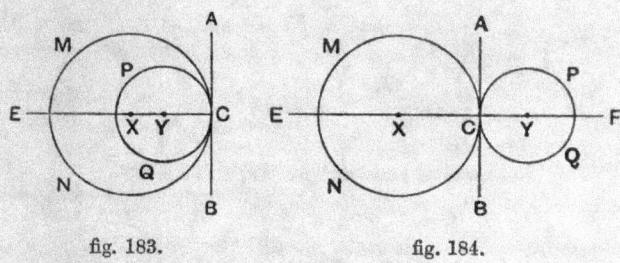

fig. 183.   fig. 184.

Data   The ⊙s **CMN, CPQ** touch internally (fig. 183) or externally (fig. 184) at **C**.

> **X, Y** are the centres of the ⊙s.
>
> **AB** is the common tangent at **C**.

To prove that   **XY** produced (fig. 183), or **XY** (fig. 184) passes through **C**.

Construction   Join **XC, YC**.

Proof   Since **CA** is the tangent at **C** to ⊙**CMN**, and **CX** the radius through **C**,

> $\therefore$ ∠ **XCA** is a rt. ∠.   III. 6.
>
> Sim^ly ∠ **YCA** is a rt. ∠.

$\therefore$ if the ⊙s touch internally, **XYC** is a straight line,

and if the ⊙s touch externally, ∠ **XCA** + ∠ **YCA** = 2 rt. ∠ s.

> $\therefore$ **XCY** is a straight line.   I. 2.

Q. E. D.

Cor.   If two circles touch externally the distance between their centres is equal to the sum of their radii; if they touch internally the distance between their centres is equal to the difference of their radii.

¶Ex. **985.** Describe the relative position of the two circles in each of the following cases (d is the distance between the centres, R and r are the radii). Do this, if you can, without drawing the circles.

(i) $d = 4\cdot1$ cm., $R = 2\cdot1$ cm., $r = 1\cdot4$ cm.

(ii) $d = 0\cdot7$ cm., $R = 2\cdot2$ cm., $r = 1\cdot2$ cm.

(iii) $d = 3\cdot4$ cm., $R = 2\cdot0$ cm., $r = 1\cdot4$ cm.

(iv) $d = 0\cdot8$ cm., $R = 2\cdot1$ cm., $r = 1\cdot3$ cm.

(v) $d = 0$ cm., $R = 1\cdot9$ cm., $r = 1\cdot2$ cm.

(vi) $d = 1\cdot5$ cm., $R = 2\cdot0$ cm., $r = 1\cdot5$ cm.

(vii) $d = 2\cdot5$ cm., $R = 1\cdot7$ cm., $r = 1\cdot7$ cm.

Ex. **986.** What is the distance between the centres of two circles of radii 15 and 14 in. (i) if they have external contact, (ii) if they have internal contact?

Ex. **987.** Show how to draw three circles having for centres the vertices of an equilateral triangle of side 2 in., so that each circle may touch the two others externally.

Ex. **988.** Three circles, of radii 1, 2, 3 in., touch externally, each circle touching the other two. What are the distances between the centres? Draw the circles.

CONSTRUCTION OF CIRCLES TO SATISFY GIVEN CONDITIONS.

¶Ex. **989. What is the locus of the centres of all circles of radius 1 in., which touch externally a fixed circle of radius 2 in.? Draw the locus, and draw a number of the touching circles.**

¶Ex. **990.** If required to draw a circle to touch a given circle *at a given point*, where would you look for the centre of the touching circle? **What is the locus of the centres of circles touching a given circle at a given point?** Draw a number of such circles, some enclosing the given circle, some inside it, some external to it.

¶Ex. **991.** **What is the locus of the centres of circles which touch a given line at a given point?**

¶Ex. **992.** What is the locus of the centres of circles of radius 1 in., touching a given circle of radius 2 in., and lying inside it? Draw a number of such circles.

¶Ex. **993.** Repeat Ex. 992 with 1 in. radius for the touching circles, and 3 in. radius for the fixed circle.

¶Ex. **994.** Draw a number of circles of radius 3 in. to touch a circle of radius 2 in. and enclose it.

¶Ex. **995.** Draw a number of circles of radius 4 in. to touch a given circle of radius 2 in. and enclose it.

¶Ex. **996.** **What is the locus of centres of circles of given radius passing through a given point?**

¶Ex. **997.** What is the locus of centres of circles (i) passing through two given points, (ii) touching two given lines?

Each of the following problems is to be solved by finding the centre of the required circle (generally by the intersection of loci). Some of the group have been solved already; they are recapitulated below for the sake of completeness. In several cases a numerical instance is given which should be attempted first, the radius of the resulting circle being measured.

Ex. **998.** Draw a circle (or circles) to satisfy the following conditions:

(i) To pass through three given points (solved already).

(ii) Of given radius, to pass through two given points (solved already).

(iii) Of given radius, to pass through a given point and touch a given line, e.g. take radius 2 in. and a point distant 1 in. from the line. (What is the locus of centres of 2 in. circles passing through given point? touching given line?) When is the general problem impossible?

(iv) To touch a given line AB at a given point P, and to pass through a given point Q outside the line. (What is the locus of centres of ⊙^{s} touching line at P? passing through P and Q? Let PQ = 3 cm., ∠QPA = 30°.)

(v) To touch a given circle at a given point P, and to pass through a given point Q not on the circle. In what case is this impossible?

(vi) To touch a given line AB at P, and also to touch a given line CD, not parallel to AB. (What is the locus of centres of circles touching AB and CD?)

(vii) Of given radius, to pass through a given point P and touch a given circle, e.g. let given radius=4 cm., radius of given circle=3 cm., distance of P from centre of given circle=5 cm. (Compare (iii).)

(viii) Of given radius, to touch a given circle at a given point (how many solutions are there?).

(ix) To touch three given lines (solved already).

(x) Of given radius to touch two given lines, e.g. let the lines intersect at an angle of 60°, and radius=1 in. (How many solutions are there?)

(xi) Of given radius, to touch a given line and a given circle (e.g. given radius=3 cm., radius of given circle=5 cm., distance of line from centre of circle=6 cm.). What is the condition that the general problem may be possible?

(xii) To touch three equal circles (a) so as to enclose them all, (b) so as to enclose none of them. (Begin by drawing a circle through the three centres.)

(xiii) Of given radius, to touch two given circles (e.g. let given radius=2 in., radii of given circles=1 in., 1·5 in., distance between centres =3·5 in.).

Ex. 999. In a semicircle of radius 5 cm. inscribe a circle of radius 2 cm. Measure the parts into which the diameter of the semicircle is divided by the point of contact.

Ex. 1000. Draw four circles of radius 2 in., touching a fixed circle of radius 1 in., and also touching a straight line 2 in. distant from the centre of the fixed circle.

fig. 185.

Ex. 1001. Show how to inscribe a circle in a sector of 80° of a circle whose radius is 4 in.

Ex. 1002. Show how to draw three equal circles, each touching the other two; and how to circumscribe a fourth circle round the other three.

†Ex. **1003.** Prove that, if circles are described with centres A, B, C (fig. 180) and radii AY, BZ, CX, the three circles touch.

†Ex. **1004.** A variable circle (centre O) touches externally each of two fixed circles (centres A, B). Prove that the difference of AO, BO remains constant.

†Ex. **1005.** If two circles touch and a line is drawn through the point of contact to meet the circles again at P and Q, the tangents at P and Q are parallel. (Draw the common tangent at the point of contact.)

†Ex. **1006.** If two circles touch externally at A and are touched at P, Q by a line PQ, then PQ subtends a right angle at A. Also PQ is bisected by the common tangent at A.

†Ex. **1007.** Prove that, in Ex. 1006, the circle on PQ as diameter passes through A and touches the line of centres.

†Ex. **1008. Two circles intersect at A, B; prove that the line of centres bisects AB (the common chord) at right angles.** (See III. 1 Cor.)
What axis of symmetry has the above figure?
What does this theorem become, in the limit, when the two circles touch?

Ex. **1009.** Find the distance between the centres of two circles, their radii being 5 and 7 cm. and their common chord 8 cm. (There are two cases.)

SECTION VI. ANGLE PROPERTIES.

Reflex angles. Take your dividers and open them slowly. The angle between the legs is first an acute angle, then a right angle, then an obtuse angle. When the dividers are opened out flat, the angle has become two right angles (180°). If the dividers are opened still further the angle of opening is greater than 180° and is called a **reflex** angle.

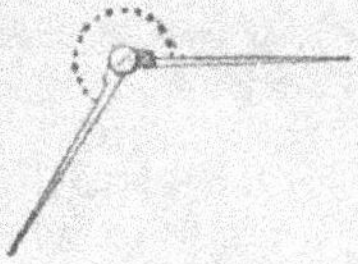

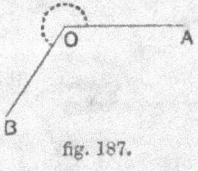

fig. 186.

DEF. A **reflex** angle is an angle greater than two right angles and less than four right angles. Fig. 187 shows two straight lines OA, OB forming a reflex angle (marked), and also an obtuse angle (unmarked).

fig. 187.

12—2

¶Ex. **1010**. Account for the necessity of the phrase "less than four right angles" in the above definition.

¶Ex. **1011**. Open a book to form a reflex angle.

¶Ex. **1012**. What is the sum of the reflex angle a and the acute angle b in fig. 188? If $\angle b = 36°$, what is $\angle a$?

¶Ex. **1013**. What kind of angle is subtended at the centre of a circle by a major arc?

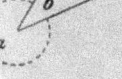

¶Ex. **1014**. Draw a quadrilateral having one angle reflex. Prove that the sum of the four angles is 360°.

fig. 188.

¶Ex. **1015**. Is it possible for (i) a four-sided figure, (ii) a five-sided figure to have *two* of its angles reflex?

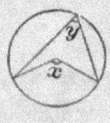   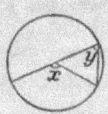   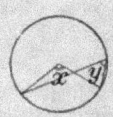

fig. 189.   fig. 190.   fig. 191.

¶Ex. **1016**. Draw a figure like fig. 189, making the radius of the circle about 2 in. Measure angles x and y. Notice that they are subtended by the same arc.

¶Ex. **1017**. Do the same for figs. 190, 191, 192. What relation do you notice between the angle x and the angle y in the four experiments?

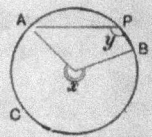

¶Ex. **1018**. Draw a circle of radius 5 cm. : place in it a chord AB of length 9·5 cm. Mark four points P, Q, R, S in the major arc. What relation do you notice between the angles subtended at these points by AB?

fig. 192.

¶Ex. **1019**. In the figure of Ex. 1018 mark three points X, Y, Z in the minor arc. What do you notice about the angles subtended at X, Y, Z by AB?

¶Ex. **1020**. Draw a circle and a diameter. Mark four points on the circle, at random. What are the angles subtended by the diameter at each of these points?

†Ex. **1021**. A side BA of an isosceles triangle ABC is produced, through the vertex A, to a point D. Prove that $\angle DAC = 2\angle ABC = 2\angle ACB$.

Theorem 8.

The angle which an arc of a circle subtends at the centre is double that which it subtends at any point on the remaining part of the circumference.

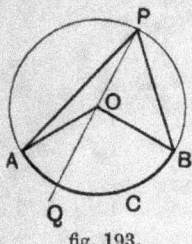 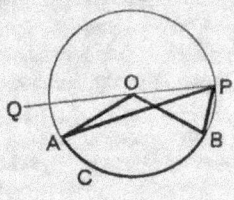

fig. 193.   fig. 194.

Data   The arc ACB of ⊙ ACB subtends ∠ AOB at the centre O; and subtends ∠ APB at P, any point on the remaining part of the circumference.

To prove that   ∠ AOB = 2 ∠ APB.

Construction   Join PO, and produce to Q.

Proof CASE I.   *When the centre* O *is inside* ∠ APB.
                In △ AOP,   OA = OP (radii),
                        ∴ ∠ OPA = ∠ OAP.   I. 12.
                Now ∠ QOA is an exterior ∠ of △ AOP,
                        ∴ ∠ QOA = ∠ OPA + ∠ OAP   I. 8, *Cor.* 1.
                            = 2 ∠ OPA.
                Sim^ly ∠ QOB = 2 ∠ OPB,
            ∴ ∠ QOA + ∠ QOB = 2 (∠ OPA + ∠ OPB),
                        ∴ ∠ AOB = 2 ∠ APB.

CASE II.   *When the centre* O *is outside* ∠ APB.
                As before, ∠ QOB = 2 ∠ OPB,
                and ∠ QOA = 2 ∠ OPA,
            ∴ ∠ QOB − ∠ QOA = 2 (∠ OPB − ∠ OPA),
                        ∴ ∠ AOB = 2 ∠ APB.   Q. E. D.

†Ex. **1022.** Prove the above theorem for the case in which ACB is a major arc, and the angle subtended at the centre a reflex angle (see fig. 192). What kind of angle is ∠APB in this case?

†Ex. **1023.** Prove the above theorem for the case in which O lies on AP (see fig. 190).

†Ex. **1024.** Prove that in fig. 195 ∠a=∠b.

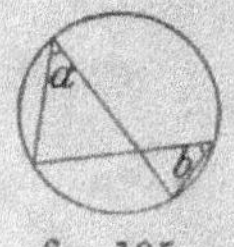

Ex. **1025.** **Show how to inscribe in a given circle a triangle of given angles.** (Consider what angles the sides subtend at the centre.)

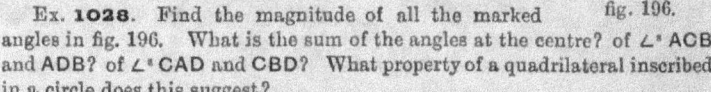

fig. 195.

†Ex. **1026.** If the two circles in figs. 193 and 194 are equal, and the arcs ACB are equal, prove that the angles APB are equal.

Ex. **1027.** Draw a freehand figure for the case of III. 8 in which arc ACB is a semicircle. What does ∠AOB become in this case? What does ∠APB become?

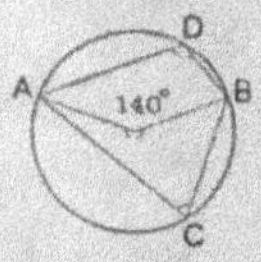

fig. 196.

Ex. **1028.** Find the magnitude of all the marked angles in fig. 196. What is the sum of the angles at the centre? of ∠ˢ ACB and ADB? of ∠ˢ CAD and CBD? What property of a quadrilateral inscribed in a circle does this suggest?

DEF. A **segment** of a circle is the part of the plane bounded by an arc and its chord.

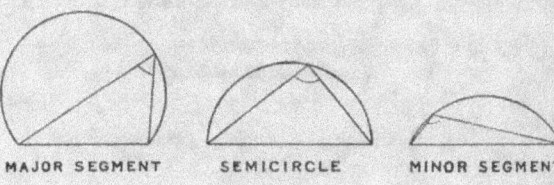

MAJOR SEGMENT   SEMICIRCLE   MINOR SEGMENT
fig. 197.

DEF. An **angle in a segment** of a circle is an angle subtended by the chord of the segment at a point on the arc (fig. 197).

DEF. A segment is called a **major** segment or a **minor** segment according as its arc is a major or a minor arc. It is obvious that a major segment of a circle is greater than the semicircle (considered as an area) and that a minor segment is less than the semicircle.

THEOREM 9.

Angles in the same segment of a circle are equal.

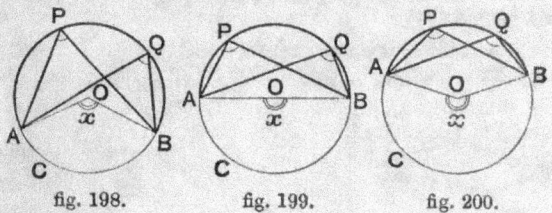

fig. 198.   fig. 199.   fig. 200.

Data   ∠s **APB, AQB** are two ∠s in the same segment **APQB** of
⊙**APB**. (Three figures are drawn, for the three cases in
which the segment >, = or < a semicircle.)

To prove that   ∠ **APB** = ∠ **AQB**.

Construction   Join **A, B** to the centre.
      Let x be the ∠ subtended at the centre by arc **ACB**.

Proof   In each figure ∠ x = 2 ∠ **APB**,
   for these angles are subtended by the same arc **ACB**.   III. 8.
               Sim^ly ∠ x = 2 ∠ **AQB**,
               ∴ ∠ **APB** = ∠ **AQB**.

Q. E. D.

NOTE. Since all the angles in a segment are equal, we may
in future speak of *the* angle in a segment when we mean the
magnitude of any angle in the segment.

¶Ex. **1029.** Are ∠s **PAQ, PBQ** in fig. 200 equal?
Give a reason.

Ex. **1030.** Copy fig. 201 (freehand) ; join **BC, BD.**
Find all the angles in your figure.

†Ex. **1031.** **AC, BD** are perpendicular diameters
of a circle, **P** is a point on the minor arc **AB**; prove that
the arcs **AD, DC, CB** each subtend 45° at **P**.

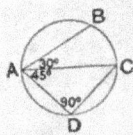

fig. 201.

Theorem 10.

The angle in a major segment is acute; the angle in a semicircle is a right angle; and the angle in a minor segment is obtuse.

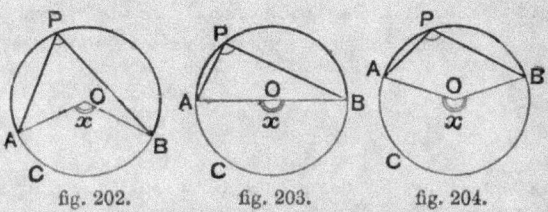

fig. 202.   fig. 203.   fig. 204.

CASE I.

Data   APB is a major segment.

To prove that $\angle$ APB is acute.

Proof   Since APB is a major segment,

          $\therefore$ arc ACB is a minor arc,

          $\therefore$ $\angle x < 2$ rt. $\angle$ s.

          But $\angle$ APB $= \frac{1}{2} \angle x$.   III. 9.

          $\therefore$ $\angle$ APB < 1 rt. $\angle$.

CASE II.

Data   APB is a semicircle.

To prove that $\angle$ APB is a rt. $\angle$.

Proof   Since APB is a semicircle, so also is ACB,

          $\therefore$ $\angle x = 2$ rt. $\angle$ s,

          $\therefore$ $\angle$ APB $= 1$ rt. $\angle$.

CASE III.

Data   APB is a minor segment.

To prove that $\angle$ APB is obtuse.

Proof   Since APB is a minor segment,

          $\therefore$ arc ACB is a major arc,

          $\therefore$ $\angle x > 2$ rt. $\angle$ s,

          $\therefore$ $\angle$ APB > 1 rt. $\angle$.

                                Q. E. D.

Ex. 1032. A regular hexagon is inscribed in a circle. What is the angle in each of the segments of the circle which lie outside the hexagon?

Ex. 1033. Repeat Ex. 1032 for the case of (i) a square, (ii) an equilateral △, (iii) a regular *n*-gon.

†**Ex. 1034.** A, B, C, D are points on a circle; the diagonals of ABCD meet at X; prove that △ˢ ABX, DCX are equiangular; as also △ˢ BCX, ADX.

†**Ex. 1035.** Through X, a point outside a circle, XAB, XCD are drawn to cut the circle in A, B; C, D. Prove that △ˢ XAD, XCB are equiangular.

†**Ex. 1036.** Prove the following construction for erecting a perpendicular to a given straight line AB at its extremity B. With centres A, B describe arcs of equal circles, cutting at C. With centre C and radius CA describe a circle. Produce AC to meet this circle again in D; then BD is ⊥ to AB.

†**Ex. 1037.** The circle described on a side of an isosceles triangle as diameter bisects the base.

†**Ex. 1038.** The circles drawn on two sides of a triangle as diameters intersect on the base.

†**Ex. 1039.** The four circles drawn with the sides of a rhombus for diameters have one point in common.

†**Ex. 1040.** Two circles intersect at P, Q. Through P diameters PA, PB of the two circles are drawn. Show that AQ, QB are in the same straight line. (Join QP.)

†**Ex. 1041.** AD is ⊥ to the base BC of △ABC; AE is a diameter of the circumscribing circle. Prove that △ˢ ABD, AEC are equiangular; as also △ˢ ACD, AEB.

†**Ex. 1042.** The bisector of A, the vertical angle of △ABC, meets the base in D and the circumscribing circle in E. Prove that △ˢ ABD, AEC are equiangular. Also prove that △ˢ ACD, AEB are equiangular.

THEOREM 11.

[CONVERSE OF THEOREM 9.]

If the line joining two points subtends equal angles at two other points on the same side of it, the four points lie on a circle.

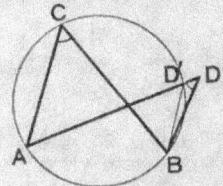 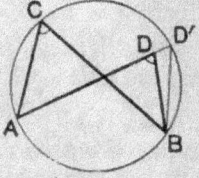

fig. 205.   fig. 206.

Data   The line joining AB subtends equal ∠s at the points C, D, which lie on the same side of AB.

To prove that   the four points A, B, C, D lie on a ⊙.

Construction   Draw ⊙ to pass through A, B and C.

It must be shown that this ⊙ passes through D.

Proof   If ⊙ABC does not pass through D, it must cut AD (or AD produced) in some other point D'.

Join BD'.

Then ∠AD'B = ∠ACB (in same segment).   III. 9.

But ∠ADB = ∠ACB,   *Data*

∴ ∠AD'B = ∠ADB.

But this is impossible, for one of the ∠s is an exterior ∠ of △DD'B, and the other is an interior opposite ∠ of the same △.

Hence ⊙ABC must pass through D,

i.e. A, B, C, D lie on a ⊙.   Q. E. D.

DEF.   Points which lie on the same circle are said to be **concyclic.**

†Ex. **1043.** BE, CF are altitudes of the triangle ABC; prove that B, F, E, C are concyclic. Sketch in the circle.

¶Ex. **1044.** Draw a circle (radius about 3 in.); take four points A, B, C, D upon it. By measurement, find the sum of the angles BAD, BCD; also of the angles ABC, ADC.

THEOREM 12.

The opposite angles of any quadrilateral inscribed in a circle are supplementary.

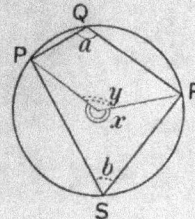
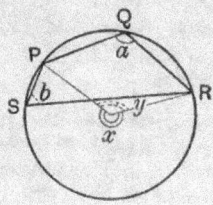

fig. 207.   fig. 208.

Data   PQRS is a quadrilateral inscribed in ⊙ PQR*.

To prove that   (1)   ∠ PQR + ∠ PSR = 2 rt. ∠ s,

              (2)   ∠ SPQ + ∠ SRQ = 2 rt. ∠ s.

Construction   Join P and R to the centre of ⊙.

Proof $\angle a = \frac{1}{2} \angle x,$ III. 8.

                           $\angle b = \frac{1}{2} \angle y,$ III. 8.

                $\therefore \angle a + \angle b = \frac{1}{2}(\angle x + \angle y).$

                But ∠ x + ∠ y = 4 rt. ∠ s,

                $\therefore \angle a + \angle b = 2$ rt. ∠ s,

                i.e. ∠ PQR + ∠ PSR = 2 rt. ∠ s.

Sim^ly it may be shown that ∠ SPQ + ∠ SRQ = 2 rt. ∠ s.

Q. E. D.

* The two figures represent the two cases in which the centre is (i) inside, (ii) outside the quadrilateral. The same proof applies to both.

Cor. **If a side of a quadrilateral inscribed in a circle is produced, the exterior angle so formed is equal to the interior opposite angle of the quadrilateral.**

†Ex. **1045.** Prove the corollary.

¶Ex. **1046.** What is the relation between the angles subtended by a chord at a point in its minor arc, and at a point in its major arc?

Ex. **1047.** From the given angles, find all the angles in fig. 209.

Ex. **1048.** Repeat Ex. 1047, taking ∠B=71°, ∠BCO=36°, ∠AOD=108°. Prove that in this case AD is ∥ to BC.

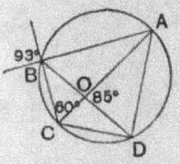

fig. 209.

Ex. **1049. Show how to circumscribe about a given circle a triangle of given angles.** (Suppose the figure drawn, and the points of contact joined to the centre: what are the angles at the centre?)

†Ex. **1050.** If a parallelogram can be inscribed in a circle, it must be a rectangle.

†Ex. **1051.** If a trapezium can be inscribed in a circle, it must be isosceles.

†Ex. **1052.** The sides BA, CD of a quadrilateral ABCD, inscribed in a circle, are produced to meet at O; prove that △ˢ OAD, OCB are equiangular.

†Ex. **1053.** ABCD is a quadrilateral inscribed in a circle, having ∠A=60°; O is the centre of the circle. Prove that

$$∠OBD + ∠ODB = ∠CBD + ∠CDB.$$

†Ex. **1054.** ABC is an isosceles triangle, D is any point in the base BC. Prove that the circumcircles of the triangles ABD, ACD are equal.

THEOREM 13.

[CONVERSE OF THEOREM 12.]

If a pair of opposite angles of a quadrilateral are supplementary, its vertices are concyclic.

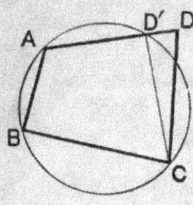 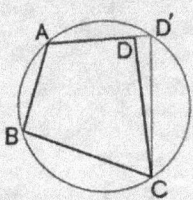

fig. 210.   fig. 211.

Data The ∠s ABC, ADC of the quadrilateral ABCD are supplementary.

To prove that A, B, C, D are concyclic.

Construction Draw ⊙ to pass through A, B, C.
It must be shown that this ⊙ passes through D.

Proof If ⊙ABC does not pass through D, it must cut AD (or AD produced) in some other point D'.
Join CD'.
Then ∠AD'C + ∠ABC = 2 rt. ∠s.   III. 12.
But ∠ADC + ∠ABC = 2 rt. ∠s,   *Data*
∴ ∠AD'C + ∠ABC = ∠ADC + ∠ABC,
∴ ∠AD'C = ∠ADC.

But this is impossible, for one of the ∠s is an exterior ∠ of △DD'C, and the other is an interior opposite ∠ of the same △.
Hence ⊙ABC must pass through D,
i.e. A, B, C, D are concyclic.

Q. E. D.

DEF. If a quadrilateral is such that a circle can be circum-scribed round it, the quadrilateral is said to be **cyclic**.

†Ex. **1055**. BE, CF, two altitudes of △ABC, intersect at H. Prove that AEHF is a cyclic quadrilateral. Sketch in the circle.

Ex. **1056**. ABC, DBC are two congruent triangles on opposite sides of the base BC. Under what circumstances are A, B, C, D concyclic?

†Ex. **1057**. ABCD is a parallelogram. A circle drawn through A, B, cuts AD, BC (produced if necessary) in E, F respectively. Prove that E, F, C, D are concyclic.

†Ex. **1058**. ABCD is a quadrilateral inscribed in a circle. DA, CB are produced to meet at E; AB, DC to meet at F. Prove that, if a circle can be drawn through AEFC, then EF is a diameter of this circle; and BD is the diameter of ⊙ABCD.

†Ex. **1059**. The straight lines bisecting the angles of any convex quadrilateral form a cyclic quadrilateral.

¶Ex. **1060**. Stick two pins into the paper 2 in. apart at A and B; place the set-square on the paper so that the sides containing the 60° are in contact with the pins; mark the point where the vertex of the angle rests. Now slide the set-square about, keeping the same two sides against the pins, and plot the locus of the 60° vertex. What is the locus? are A, B points in the locus?

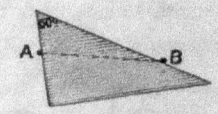

fig. 212.

¶Ex. **1061**. Repeat the experiment of Ex. 1060 with the 30° vertex.

¶Ex. **1062**. Repeat the experiment of Ex. 1060 with the 90° vertex.

¶Ex. **1063**. (Tracing paper.) Draw, on tracing paper, two straight lines intersecting at P. On your drawing paper mark two points A, B. Move your tracing paper about so that the one line may always pass through A, and the other through B. Plot the locus of P by pricking through.

The foregoing exercises will have prepared the reader for the following statement:

The locus of points (on one side of a given straight line) at which the line subtends a constant angle is an arc of a circle, the given line being the chord of the arc.

¶Ex. **1064.** Upon what theorem does the truth of the above statement depend?

¶Ex. **1065.** What kind of arc is obtained if the constant angle is (i) acute, (ii) a right angle, (iii) obtuse?

¶Ex. **1066.** If the constant angle is 45°, what angle is subtended by the given line at the centre of the circle? Using this suggestion, show how to draw the locus of points at which a line of 5 cm. subtends 45°, without actually determining any of the points.

†Ex. **1067.** Show how to construct the locus of points at which a given line subtends a given angle.

Ex. **1068.** On a chord of 3·5 in. construct a segment of a circle to contain an angle of 70°. Measure the radius.

Ex. **1069.** Repeat Ex. 1068 with chord of 7·24 cm. and angle of 110°.

†Ex. **1070.** Prove that the locus of the mid-points of chords of a circle which are drawn through a fixed point is a circle.

†Ex. **1071.** Of all triangles of given base and vertical angle, the isosceles triangle has greatest area.

†Ex. **1072.** P is a variable point on an arc AB. AP is produced to Q so that PQ = PB. Prove that the locus of Q is a circular arc.

To construct a triangle with given base, given altitude, and given vertical angle.

Let the base be 7 cm.; the altitude 6·5 cm.; the vertical angle 46°.

Draw the given base.

Draw the locus of points at which the given base subtends 46°.

Draw the locus of points distant 6·5 cm. from the given base (produced if necessary).

The intersections of these loci will be the required positions of the vertex.

How many solutions are there to this problem?

Measure the base angles of the triangle.

Ex. **1073.** Construct a triangle having

    (i)   base = 4 in.,   altitude = 1 in.,   vertical angle = 90°.

    (ii)   base = 10 cm., altitude = 2 cm.,   vertical angle = 120°.

    (iii)   base = 8 cm.,   altitude = 5 cm.,   vertical angle = 90°.

    (iv)   base = 3·5 in., altitude = 1 in.,   vertical angle = 54°.

In each case measure the base angles.

Ex. **1074.** Show how to construct a triangle given (i) base, vertical angle, and median, (ii) base, vertical angle, and area.

Ex. **1075.** Show how to construct a quadrilateral ABCD, given AB, AC, AD, ∠BAD, ∠BCD.

Ex. **1076.** Show how to construct a cyclic quadrilateral ABCD, given AB, BC, CD, ∠B.

Ex. **1077.** Show how to construct a quadrilateral ABCD, given AB, BC, CA, AD, ∠BDC.

Ex. **1078.** Show how to construct a parallelogram given the base and height, and the angle (subtended by the base) at the intersection of the diagonals.

Looking at the tangent as the limit of a chord, it will be seen from fig. 213 that as B moves round towards A the constant

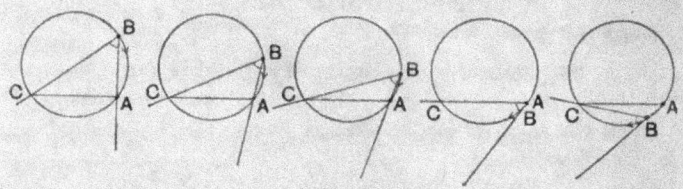

fig. 213.

angle CBA becomes, in the limit, the angle between the chord and the tangent at A.   This suggests the following theorem.

THEOREM 14.

If a straight line touch a circle, and from the point of contact a chord be drawn, the angles which this chord makes with the tangent are equal to the angles in the alternate segments.

Data   AB touches ⊙ CDE in C; the chord CD is drawn through C, meeting ⊙ again in D.

To prove that   (1)   ∠ BCD = ∠ in alternate segment CED,

(2)   ∠ ACD = ∠ in alternate segment CFD (fig. 215).

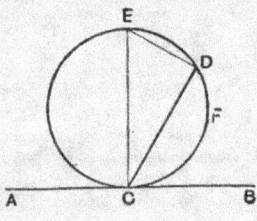

fig. 214.

(1)   *Construction*   Through C draw CE ⊥ to AB, meeting ⊙ in E.
Join CE, DE.

Proof   Since CE is drawn ⊥ to tangent AB, at its point of contact C,

∴ CE passes through centre of ⊙, and is a diameter,

III. 6, *Cor.*

∴ ∠ CDE is a rt. ∠,   III. 10.

∴ in △ CDE, ∠ CED + ∠ DCE = 1 rt. ∠.   I. 8.

Now ∠ BCD + ∠ DCE = 1 rt. ∠.   *Constr.*

∴ ∠ BCD + ∠ DCE = ∠ CED + ∠ DCE,

∴ ∠ BCD = ∠ CED.

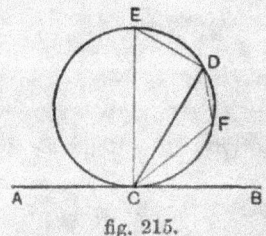

fig. 215.

(2)   *Construction*   Take any point F in arc CFD; join CF, DF.

$$\angle BCD + \angle ACD = 2 \text{ rt. } \angle s.$$

Also, since CFDE is a quadrilateral inscribed in a circle,

$$\angle CED + \angle CFD = 2 \text{ rt. } \angle s, \qquad\qquad \text{III. 12.}$$

$$\therefore \ \angle BCD + \angle ACD = \angle CED + \angle CFD.$$

But $\angle BCD = \angle CED,$ *Proved*

$$\therefore \ \angle ACD = \angle CFD.$$

Q. E. D.

¶Ex. **1079.**   In fig. 215 point out an angle equal to ∠BCF.

¶Ex. **1080.**   Taking CE as the chord (fig. 215), what is the segment alternate to ∠ACE?

Ex. **1081.**   Find all the angles of fig. 215, supposing that ∠BCD=60°, and that ∠FCD=20°. What angles do the chords ED, CD, FC subtend at the centre?

Ex. **1082.**   Find all the angles of fig. 216.

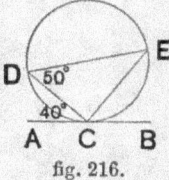

fig. 216.

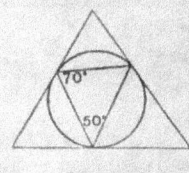

fig. 217.

Ex. **1083.**   Find all the angles of fig. 217.

Ex. **1084.**   In fig. 214 if CF is drawn to bisect ∠BCD, prove that the chords CF and FD are equal.

Ex. **1085.** If in fig. 214 arc CD=2 arc DE, what is ∠BCD?

Ex. **1086.** If in fig. 214 arc ED were ¼ arc DC, what would be the magnitude of ∠BCD?

Ex. **1087.** Show how to construct on a given straight line a segment of a circle to contain a given angle. (Draw the tangent at the extremity of the chord.)

Ex. **1088.** Show how to inscribe in a given circle a triangle of given angles. (Supposing the figure drawn, and the tangent at one vertex also drawn, the three angles at this point are known.)

MISCELLANEOUS EXERCISES ON ANGLE PROPERTIES OF CIRCLES.

†Ex. **1089.** Through P, Q, the points of intersection of two circles, are drawn chords APB, CQD; prove that AC is ∥ to BD. [Join PQ.]
What does this theorem become if A, C are made to coincide?

†Ex. **1090.** Through P, Q, the points of intersection of two circles, are drawn parallel chords APB, CQD; prove that AB=CD.

†Ex. **1091.** If two opposite sides of a cyclic quadrilateral are equal, the other two are parallel.

†Ex. **1092.** Each of two equal circles passes through the centre of the other: AB is their common chord. Through A is drawn a line cutting the two circles again in C, D; prove that △BCD is equilateral.

†Ex. **1093.** ABC is an equilateral triangle inscribed in a circle; P is any point on the minor arc BC. Prove that PA=PB+PC. [Make PX=PB. Then prove XA=PC.]

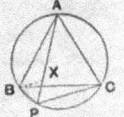

†Ex. **1094.** In fig. 181, B, C, I₁, and the centre of the inscribed circle are concyclic.

fig. 218.

Ex. **1095.** What theorem is suggested by the last figure of fig. 213?

†Ex. **1096.** From a point on the diagonal of a square, lines PR, QS are drawn parallel to the sides, P, Q, R, S being on the sides. Prove that these four points are concyclic.

†Ex. **1097.** O is the centre of a circle, CD a diameter, and AB a chord perpendicular to CD. If B is joined to any point E in CD, and BE produced to meet the circle again in F, then A, O, E, F are concyclic.

Ex. **1098**. Show how to construct a right-angled triangle, given the radius of the inscribed circle, and an acute angle of the triangle.

†Ex. **1099**. Two circles touch at A.   Through A are drawn straight lines PAQ, RAS; cutting the circles in P, Q and R, S.   Prove that PR is parallel to QS.   (Draw tangent at A.   Compare Ex. 1089.)

†Ex. **1100**. Two circles cut at P, Q.   A, a point on the one circle, is joined to P, Q; and these lines are produced to meet the other circle in B, C.   Prove that BC is parallel to the tangent at A.   (Compare Ex. 1089.)

†Ex. **1101**. A chord AB of a circle bisects the angle between the diameter through A, and the perpendicular from A upon the tangent at B.

†Ex. **1102**. ABCD is a cyclic quadrilateral, whose diagonals intersect at E: a circle is drawn through A, B and E.   Prove that the tangent to this circle at E is parallel to CD.

†Ex. **1103**. AB, AC are two chords of a circle; BD is drawn parallel to the tangent at A, to meet AC in D; prove that ∠ABD is equal or supplementary to ∠BCD.   Hence show that the circle through B, C and D touches AB at B.

†Ex. **1104**. **Prove that equal arcs or chords of a circle subtend equal (or supplementary) angles at a point on the circumference.** Make a sketch to illustrate the case of supplementary angles.

NOTE. *In the following exercises* (*Exs*. 1105—1118) *the student is advised to make use of* "*the angle subtended at the circumference.*"

†Ex. **1105**. ABCDE is a regular pentagon inscribed in a circle.

(i)   Prove that the angle A is trisected by AC, AD.

(ii)   Prove that AB is parallel to EC.   (Join AC.)

(iii)   At what angle do BD, CE, intersect?

(iv)   Prove that △ ACD is isosceles, and that each of its base angles is double its vertical angle.

(v)   If BD, CE meet at X, prove that △ ˢCXD, CDE are equiangular.

(vi)   Prove that the tangent to the circle at A is parallel to BE. [Use III. 14.]

†Ex. **1106**. AB, CD are parallel chords of a circle.   Prove that arc AC = arc BD.

†Ex. **1107.** On a circle are marked off equal arcs AC, BD. Prove that AD is parallel or equal to CB.

†Ex. **1108.** AOB, COD are two chords of a circle, intersecting at right angles. Show that arc AC + arc BD = arc CB + arc DA.

†Ex. **1109.** If P is a given point inside a circle, show how to draw through P a chord of the circle cutting off the least possible minor segment.

Ex. **1110.** In fig. 219 what fractions of the circumference are the arcs AB, BC, CD, DA, BCD?

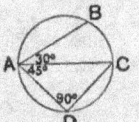

fig. 219.

Ex. **1111.** In fig. 220 what fractions of the circumference are the arcs AB, BC, CD, DA?

Ex. **1112.** ABC and ADEFG are respectively an equilateral triangle and a regular pentagon inscribed in a circle. What fraction of the circumference is the arc BD?

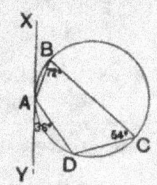

Ex. **1113.** PQRS is a quadrilateral inscribed in a circle; the two diagonals intersect at A. PQ is an arc of 30° (see p. 162), QR 100°, RS 70°. Find all the angles in the figure.

fig. 220.

Ex. **1114.** In the figure of Ex. 1113 find two pairs of equiangular triangles.

Ex. **1115.** The two tangents OA, OB from a point O are inclined at an angle of 48°. How many degrees are there in the minor and major arc AB respectively? What is the ratio of the major to the minor arc?

†Ex. **1116.** P is a variable point on an arc AB. Prove that the bisector of ∠APB always passes through a fixed point.

[Begin by finding the probable position of the fixed point by experiment.]

†Ex. **1117.** A, B, C are three points on a circle. The bisector of ∠ABC meets the circle again at D. DE is drawn ∥ to AB and meets the circle again at E. Prove that DE = BC.

†Ex. **1118.** A tangent is drawn at one end of an arc; and from the mid-point of the arc perpendiculars are drawn to the tangent, and the chord of the arc. Prove that they are equal.

Section VII. Construction of Tangents.
(Ruler and Compass.)

¶Ex. **1119.** What is the locus of points at which a given line subtends a right angle?

¶Ex. **1120.** O is the centre of a circle and Q is a point outside the circle. Construct the locus of points at which OQ subtends a right angle. Find two points A, B on the first circle, so that $\angle OAQ = \angle OBQ = 90°$. Prove that QA is a tangent to the first circle.

To draw tangents to a given circle ABC from a given point T outside the circle.

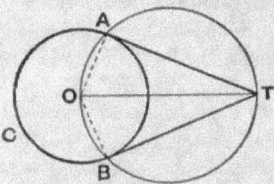

fig. 221.

Construction   Join T to O, the centre of ⊙ABC.

On OT as diameter describe a ⊙ cutting the given circle in A, B.

Join TA, TB.

These lines are tangents.

Proof   Join OA, OB.

Since OT is the diameter of ⊙OAT,

∴ ∠OAT is a right angle,

∴ AT, being ⊥ to radius OA, is the tangent at A.

Similarly BT is the tangent at B.

Ex. **1121.** Draw tangents to a circle of radius 2 in. from a point 1 in. outside the circle; calculate and measure the length of the tangents.

Ex. **1122.** Find the angle between the tangents to a circle from a point whose distance from the centre is equal to a diameter.

COMMON TANGENTS TO TWO CIRCLES.

DEF. A straight line which touches two circles is called a **common tangent** to the two circles.

Fig. 222 shows that when the circles do not intersect there are four common tangents.

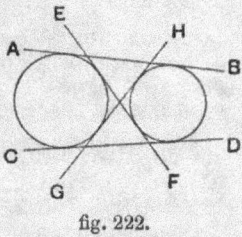

If the two circles lie on the same side of a common tangent, it is called an **exterior** common tangent; thus AB, CD (fig. 222) are exterior common tangents. If the two circles lie on opposite sides of a common tangent, it is called an **interior** common tangent; thus EF, GH are interior common tangents.

fig. 222.

Ex. **1123**. Draw sketches to show how many common tangents can be drawn in cases II., III., IV., V., of fig. 182; in each case state the number of exterior and of interior common tangents.

To construct an exterior common tangent to two unequal circles.

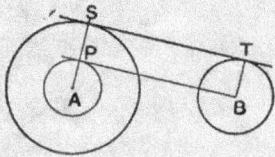

fig. 223.

[*Analysis* Let A, B be the centres of the larger and smaller circles respectively; R, r their radii.

Suppose that ST is an exterior common tangent, touching the ⊙s at S, T.

Join AS, BT. Then ∠s AST, BTS are right angles,

∴ AS is ∥ to BT.

Through B draw BP ‖ to TS, meeting AS in P.

Then BTSP is a rectangle. (Why?)

$$\therefore PS = BT = r.$$

And $AP = AS - PS = R - r$.

Also ∠ APB is a right angle. (Why?)

∴ BP is a tangent from B to a circle round A, whose radius is $R - r$.

The foregoing analysis suggests the following construction.]

Construction   With centre A describe a circle having for radius the *difference* of the given radii.

From B draw a tangent BP to this circle.

Join AP and produce it to meet the larger ⊙ in S.

Through B draw BT ‖ to AS to meet the smaller circle in T.

Join ST.

Then this line is a common tangent to the two ⊙ˢ.

Proof   PS is equal and ‖ to BT   (why?),

∴ STBP is a parallelogram,

and ∠ SPB is a right angle   (why?),

∴ STBP is a rectangle,

∴ angles at S and T are right angles,

∴ ST is a tangent to each circle.

Ex. **1124.** Draw two circles of radii 1·5 in. and 0·5 in., the centres 2·5 in. apart. Draw the two exterior common tangents.

Measure and calculate the length of these tangents (i.e. the distance between the points of contact). [Use right-angled △ APB.]

†Ex. **1125.** Where does the above method fail when the two circles are equal? Give a construction (with proof) for the exterior common tangents in this case.

To construct an interior common tangent to two circles.

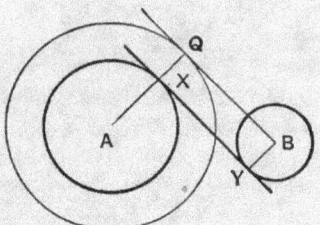

fig. 224.

[*Analysis* Suppose that XY is an interior common tangent, touching the ⊙ˢ at X, Y.

Join AX, BY. Then ∠ˢAXY, BYX are right angles.

∴ AX is ‖ to BY.

Through B draw BQ ‖ YX, meeting AX produced in Q.

Then BYXQ is a rectangle.

∴ QX = BY = r.

And AQ = AX + XQ = R + r.

Also ∠AQB is a right angle,

∴ BQ is a tangent from B to a circle round A, whose radius is R + r.

Hence the following construction.]

Construction With centre A describe a circle having for radius the *sum* of the given radii.

From B draw a tangent BQ to this circle.

Join AQ; let this line cut the (A) circle in X.

Through B draw BY ‖ QA to meet the (B) circle in Y.

Join XY.

Then this line is a common tangent to the two ⊙ˢ.

Proof (i) Prove that BYXQ is a rectangle.

(ii) Prove that XY is a tangent to the (A) circle at X, and to the (B) circle at Y.

Ex. **1126.** Draw the two circles of Ex. 1124, and draw the interior common tangents. Measure and calculate the length of these tangents.

¶Ex. **1127.** Can you suggest an easier method of constructing the internal common tangents to two equal circles?

¶Ex. **1128.** If the radius of the smaller circle diminishes till the circle becomes a point, what becomes of the four common tangents?

Ex. **1129.** The diameters of the wheels of an old-fashioned bicycle are 4 ft. and 1 ft., and the distance between the points where the wheels touch the ground is 2½ ft. Calculate the distance between the centres of the wheels.

SECTION VIII. AREA OF CIRCLE.

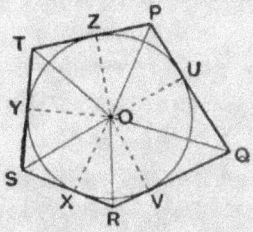

fig. 225.

Let PQRST be a polygon (not necessarily regular) circumscribing a circle.

Join the vertices of the polygon to the centre of the circle. The circle is thus divided into a number of triangles, having for bases the sides of the polygon, and for vertex the centre of the circle.

Draw perpendiculars from the centre to the sides of the polygon. These meet the sides at their points of contact and are radii of the circle. Thus the triangles OPQ, OQR, etc. are all of height equal to the radius of the circle.

Let r be the radius of the circle, a, b, c, d, e the sides of the polygon ($PQ = a$, $QR = b$, etc.).

The area of $\triangle OPQ$ is $\frac{1}{2} ar$; $\triangle OQR = \frac{1}{2} br$, etc.

$\therefore$ area of polygon $= \frac{1}{2} ar + \frac{1}{2} br + \frac{1}{2} cr + \frac{1}{2} dr + \frac{1}{2} er$

$$= \frac{1}{2} r (a + b + c + d + e)$$

$$= \frac{1}{2} \text{ radius} \times \text{perimeter of polygon.}$$

This is true for any polygon circumscribing the circle.

If we draw a polygon of a very great number of sides, it is difficult to distinguish it from the circle itself. The area of the polygon approaches closer and closer to the area of the circle; and the perimeter of the polygon to the circumference of the circle. Hence we conclude that

area of a circle $= \frac{1}{2}$ radius $\times$ circumference of circle

$$= \frac{1}{2} r \times 2\pi r$$

$$= \pi r^2.$$

[*In the following exercises it will generally be sufficient if answers are given correct to three significant figures.*]

Ex. 1130. Calculate the area of a circle whose radius is (i) 1 inch, (ii) 2 inches.

Ex. 1131. The radius of one circle is twice the radius of another; how many times does the area of the greater contain the area of the smaller? Fig. 226 shows that the area of the greater is more than double the area of the smaller. Find the area of the shaded part of fig. 226, taking the diameter of the small circles to be 1 cm.

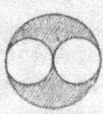

fig. 226.

Ex. 1132. Find the ratio of the area of a circle to the area of the circumscribing square.

Ex. 1133. In the centre of a circular pond of radius 100 yards is a circular island of radius 20 yards. Find the area of the surface of the water.

Ex. 1134. Find whether the area in Ex. 1133 is greater or less than the area of a circular sheet of water of 80 yards radius.

Ex. **1135**.   The radius of the inside edge of a circular running track is *a* feet; and the width of the track is *b* feet; find the area of the track.

Ex. **1136**.   From a point P, on the larger of two concentric circles, a tangent PT is drawn to the smaller.   Show that area of the circular ring between the circles is equal to that of a circle with radius PT.

Ex. **1137**.   Show how to draw a circle equal to (i) the sum, (ii) the difference of two given circles.

Ex. **1138**.   Calculate the radius of a circle whose area is 1 sq. in.

Ex. **1139**.   Calculate the diameter of a circular field whose area is 1 acre (=4840 sq. yards).

Ex. **1140**.   Prove that in fig. 172 the three portions into which the circle is divided by the curved lines are of equal area.

Ex. **1141**.   Prove that if circles are described with the hypotenuse and the two sides of a right-angled triangle for diameters, the area of the greatest is the sum of the areas of the other two.

Ex. **1142**.   In fig. 227 ∠BAC is a right angle, and the curves are semicircles.   Prove that the two shaded areas are together equivalent to the triangle.

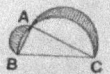

fig. 227.

Area of sector of circle.

If through the centre of a circle were drawn 360 radii making equal angles with one another, 360 angles of 1 degree would be formed at the centre of the circle.   The area of the circle would be divided into 360 equal sectors.   A sector of angle 1° has therefore $\frac{1}{360}$ of the area of the circle; and a sector of angle, say, 53° contains $\frac{53}{360}$ of the area of the circle.

Ex. **1143**.   Find the area of a sector of 40° in a circle of radius 5 in.

Ex. **1144**.   **Prove that the area of a sector of a circle is half the product of the radius and the arc of the sector.**

SECTION IX. FURTHER EXAMPLES OF LOCI.

Ex. 1145. Plot the locus of points the sum of whose distances from two fixed points remains constant.

(Mark two points S, H, say, 4 in. apart. Suppose that the point P moves so that $SP + HP = 5$ in. Then the following are among the possible pairs of values:

SP	4·5	4·0	3·5	3·0	2·5	2·0	1·5	1·0	0·5
HP	0·5	1·0	1·5	2·0	2·5	3·0	3·5	4·0	4·5

Plot all the points corresponding to all these distances, by means of intersecting arcs. Why were not values such as $SP = 4·7$, $HP = 0·3$ included in the above table? Draw a neat curve, freehand, through all these points. The locus is an oval curve called an **ellipse**.)

Ex. 1146. Describe an ellipse mechanically as follows. Stick two pins into the paper about 4 in. apart; make a loop of fine string, gut or cotton and place it round the pins (see fig. 228). Keep the loop extended by means of the point of a pencil, and move the point round the pins. It will, of course, describe an ellipse.

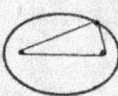

fig. 228.

Ex. 1147. Plot the locus of points the difference of whose distances from two fixed points remains constant.

(For example, let the two fixed points S, H be 4 in. apart, and let the constant difference be 2 in. Make a table as in Ex. 1145. Remember to make $SP > HP$ for some points, $HP > SP$ for other points.)

This curve is called a **hyperbola**.

Ex. 1148. Plot the locus of points the product of whose distances from two fixed points remains constant.

(For example, mark two points S, H exactly 4 in. apart. First, to plot the locus $SP \cdot HP = 5$.

Fill up the blanks in the following table:

SP	5	4·8	4	3	$\sqrt{5}$	2			
HP							3	4	4·8	5

Secondly, plot the locus $SP \cdot HP = 4$; thirdly, plot the locus $SP \cdot HP = 3$. All three loci should be drawn in the same figure.

The first locus will be found to resemble a dumb-bell, the second a figure of 8; the third consists of two separate ovals.)

Ex. **1149.** Plot the locus of a point which moves so that the ratio of its distances from two fixed points remains constant.

(For example, let the two fixed points S, H be taken 3 in. apart; and let $\dfrac{SP}{HP} = 2$.)

Ex. **1150.** OP is a variable chord passing through a fixed point O on a circle; OP is produced to Q so that PQ = OP; find the locus of Q.

Ex. **1151.** A point moves so that its distance from a fixed point S is always equal to its distance from a fixed line MN: find its locus.

(This is best done on inch paper. Take the point S 2 in. distant from the line MN. Then plot points as follows. What is the locus of points distant 3 in. from MN? distant 3 in. from S? The intersection of these two loci gives two positions of the required point. Similarly find other points.)

The curve obtained is called a **parabola**. It is the same curve as would be obtained by plotting the graph $y = \dfrac{x^2}{4} + 1$, taking for axis of x the line MN, and for axis of y the perpendicular from S to MN. It is remarkable as being the curve described by a projectile, e.g. a stone or a cricket-ball. Certain comets move in parabolic orbits, the sun being situated at the point S.

Ex. **1152.** A point moves in a plane subject to the condition that its distance from a fixed point S is always in a fixed ratio to its distance from a fixed straight line MN. Plot the curve described.

(i) Let the distance from S be always half the distance from MN. Take S 3 in. from MN.

(ii) Let the distance from S be always twice the distance from MN. Take S 3 in. from MN.

These curves will be recognised as having been obtained already.

Ex. **1153.** The ends of a rod slide on two wires which cross at right angles. Find the locus of a point on the rod. See fig. 229.

(Represent the rod by a line of 10 cm.; take the point 3 cm. from one end of the rod; also plot the locus of the mid-point. Draw the wires on your drawing paper; the rod on **tracing paper**. Keep the ends of the rod on the wires, and prick through the different positions of the point.

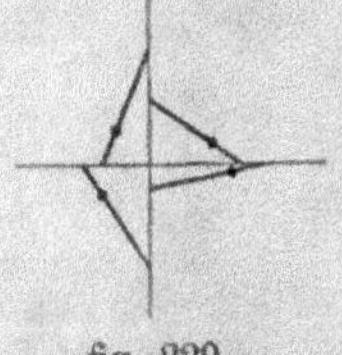

fig. 229.

Ex. **1154.** Two points A, B of a straight line move along two intersecting lines at right angles. Plot the locus of a point P, in AB produced. [Tracing paper.]

Ex. **1155.** Plot the locus of a point on the connecting-rod of a steam-engine.

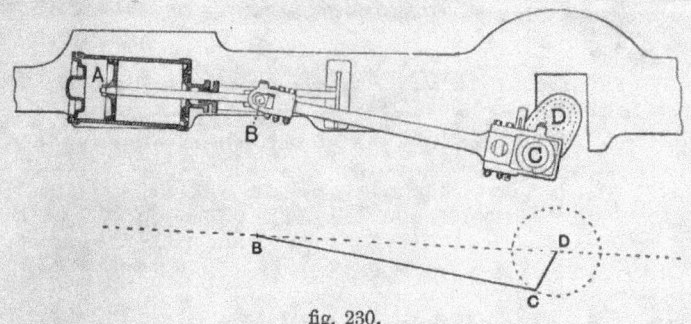

fig. 230.

(The upper diagram in fig. 230 represents the cylinder, piston-rod (AB), connecting-rod (BC), and crank (CD) of a locomotive. In the lower diagram the different parts are reduced to lines. B moves to and fro along a straight line, C moves round a circle. Take BC=10 cm., CD=3 cm. Plot the locus of a point P on BC, where BP is (i) 1 cm., (ii) 5 cm., (iii) 9 cm. This may be done, either by drawing a large number of different positions of BC; or, much more easily, by means of tracing paper.)

Ex. **1156.** A rod moves so that it always passes through a fixed point while one end always lies on a fixed circle. Plot the locus of the other end.

(Tracing paper should be used. A great variety of curves may be obtained by varying the position of point and circle, and the length of the rod. It will be seen that this exercise applies to the locus of a point on the piston-rod of an oscillating cylinder; also to the locus of a point on the stay-bar of a casement window.)

Ex. **1157.** Draw two intersecting lines. On tracing paper mark three points A, B, C. Make A slide along one line and B along the other; plot the locus of C.

Ex. **1158.** Draw two equal circles of radius 4 cm., their centres being 10 cm. apart. The two ends of a line PQ, 10 cm. in length, slide one along each circle. Plot the locus of the mid-point of PQ; also of a point 1 cm. from P.

(Most quickly done with tracing paper. It is easy to construct a model machine to describe the curve.)

Ex. **1159.** Draw two circles. On tracing paper mark three points A, B, C. Make A slide along one circle, B along another, and plot the locus of C. (Experiment with different circles and arrangements of points. You will find that in at least one case the locus-curve shrinks to a single point.)

Ex. 1160. OA, AP are two rods jointed at A. OA revolves about a hinge at O, and AP revolves twice as fast as OA, in the same direction. Find the path of a point on AP. (Make OA=2 in., AP=2½ in. Plot the locus of

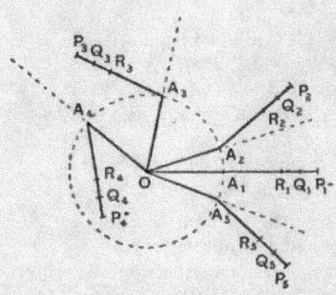

fig. 231.

P; also of Q and R, taking AQ=2 in., AR=1½ in. To draw the different positions of the rod, notice that when OA has turned through, say, 30°, AP has turned through 60° and therefore makes an angle of 30° with OA produced.)

The loci are different forms of the **limaçon**; the locus of Q is heart-shaped, and is called a **cardioid**. The locus of P has a small loop in it.

Ex. 1161. Repeat Ex. 1160, with the difference that, as OA revolves, AP remains parallel to its original position.

Ex. 1162. Two equal rods OA, AQ, jointed as in Ex. 1160, revolve in *opposite* directions at the same rate. Find the locus of Q and of the mid-point of AQ.

Ex. 1163. O is a fixed point on a circle of radius 1 in. OP, a variable chord, is produced to Q, PQ being a fixed length; also PQ'(=PQ) is marked off along PO. Plot the locus of Q and Q' when PQ is (i) 2½ in., (ii) 2 in., (iii) 1½ in.

(Draw a long line on tracing paper, and on it mark P, Q and Q'.)

Ex. 1164. Through a fixed point S is drawn a variable line SP to meet a fixed line MN in P. From P a fixed length PQ is measured off along SP (or SP produced). Find the locus of Q.

(Use tracing paper. Take S 1 in. from MN. Plot the locus of Q

   (i) when PQ=1 in., measured from P away from S,

  (ii) when PQ=1 in., measured from P towards S,

 (iii) when PQ=2 in., measured from P towards S.)

The curves obtained are different forms of the **conchoid**.

Ex. **1165.** A company of soldiers are extended in a straight line. At a given signal, they all begin to move towards a certain definite point, at the regulation pace. Are they in a straight line after 3 minutes? If not, what curve do they form?

Ex. **1166.** XOX', YOY' are two fixed straight lines, C is a fixed point (see fig. 232). A variable line PQ is drawn through C to meet XOX', YOY' in P, Q respectively. Plot the locus of the mid-point of PQ.

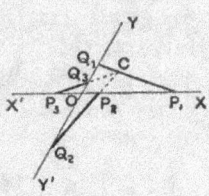

(Let XOX', YOY' intersect at 60°, and take C on the bisector of ∠XOY, 5 cm. from O.)

Ex. **1167.** (Inch paper.) Draw a circle of radius 2 in. and a straight line distant 6 in. from the centre of the circle. P is a variable point on

fig. 232.

the circle; Q is the mid-point of PN, the perpendicular from P upon the line. Plot the locus of Q.

ENVELOPES.

We have seen that a set of points, plotted in any regular way, marks out a curve which is called the locus of the points.

In a rather similar manner, a set of lines (straight or curved), drawn in any regular way, marks out a curve which is called the **envelope** of the lines. Each of the lines touches the envelope.

Let a piece of paper be cut out in the shape of a circle, and a point S marked on it. Then fold the paper so that the circumference of the circle may pass through S. If this is done many times, the creases left on the paper will envelope an ellipse (fig. 233).

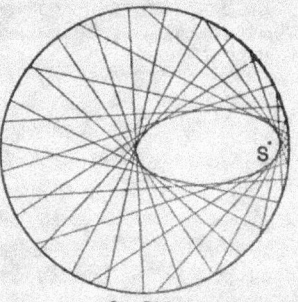

fig. 233.

Ex. **1168.** Take a piece of cardboard with one edge straight; drive a pin through the cardboard into the paper underneath; then turn the cardboard round the pin, and in each position use the straight edge of the cardboard to rule a line. What is the envelope of these lines?

Ex. 1169. One edge of a flat ruler is made to pass through a fixed point, and lines are drawn with the other edge. Find their envelope.

Ex. 1170. Prove that the envelope of straight lines which lie at a constant distance from a fixed point is a circle.

Ex. 1171. Find the envelope of equal circles whose centres lie on a fixed straight line.

Ex. 1172. Find the envelope of a set of equal circles whose centres are on a fixed circle when the radius of the equal circles is (i) less than, (ii) equal to, (iii) greater than, the radius of the fixed circle.

Ex. 1173. Draw a straight line MN and drive a pin into your paper at a point S $\frac{1}{2}$ in. from MN (see fig. 234). Keep the short edge (AB) of your set-square pressed against the pin, and keep the right angle (B) on the line MN. Rule along BC; and thus plot the envelope of BC, as the set-square slides on the paper. (Lines must of course be drawn with the set-square placed on the left of S, as well as on the right.)

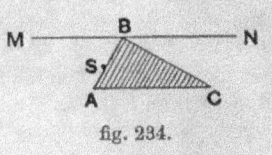

fig. 234.

Ex. 1174. Repeat Ex. 1173 using the 30° angle instead of the right angle, and putting the pin 1 in. from MN.

Ex. 1175. Draw a circle of radius 5 cm. and mark a point S 4 cm. from the centre. Let a variable line SP meet the circle in P and let PQ be drawn perpendicular to SP. Find the envelope of PQ. (The part of PQ *inside* the circle is the important part.)

Ex. 1176. Repeat Ex. 1175 with the point S on the circle.

Ex. 1177. Find the envelope of circles passing through a fixed point O, and having their centres on a fixed circle.

(i) Take radius of fixed circle = 4 cm., distance of O from centre of fixed circle = 3·2 cm.

(ii) Take radius of fixed circle = 4 cm., distance of O from centre of fixed circle = 4 cm.

(iii) Take radius of fixed circle = 3 cm., distance of O from centre of fixed circle = 5 cm.

Ex. 1178. Find the envelope of circles passing through a fixed point, and having their centres on a fixed straight line.

Ex. 1179. Plot the envelope of a straight line of constant length whose ends slide upon two fixed lines at right angles.

MISCELLANEOUS EXERCISES.

Ex. **1180**.   Show how to trisect an arc of 90°.

Ex. **1181**.   Show how to trisect a given semicircular arc.

†Ex. **1182**.   There are two fixed concentric circles; AB is a variable diameter of the one, and P a variable point on the other.   Prove that $AP^2 + BP^2$ remains constant.

[Use Apollonius' theorem, Ex. 817.]

†Ex. **1183**.   Points A, P, B, Q, C, R are taken in order on a circle so that arc AP = arc BQ = arc CR.   Prove that the triangles ABC, PQR are congruent.

Ex. **1184**.   The railway from P to Q consists of a circular arc AB and two tangents PA, BQ.   AB is an arc of 28° of a circle whose radius is $\frac{1}{2}$ mile; PA = 1 mile, BQ = $\frac{1}{2}$ mile.   Draw the railway, on a scale of 2 inches to the mile, and measure the distance from P to Q as the crow flies.   Also calculate the distance as the train goes.

†Ex. **1185**.   From a point P on a circle, a line PQ of constant length is drawn parallel to a fixed line.   Plot the locus of Q, as P moves round the circle.   Having discovered experimentally the shape of the locus, prove it theoretically.

†Ex. **1186**.   YZ is the projection of a diameter of a circle upon a chord AB; prove that AY = BZ.

†Ex. **1187**.   Through two given points P, Q on a circle draw a pair of equal and parallel chords.   Give a proof.

†Ex. **1188**.   AOB, COD are two variable chords of a circle, which are always at right angles and pass through a fixed point O.   Prove that $AB^2 + CD^2$ remains constant.

†Ex. **1189**.   Through A, a point inside a circle (centre O), is drawn a diameter BAOC; P is any point on the circle.   Prove that AC > AP > AB.

Ex. **1190**.   What is the length of (i) the shortest, (ii) the longest chord of a circle of radius r, drawn through a point distant d from the centre?

Ex. **1191**.   Two chords of a circle are at distances from its centre equal to $\frac{3}{5}$ and $\frac{4}{5}$ of its radius.   Find how many times the shorter chord is contained in three times the longer chord.

14—2

Ex. **1192.** The star-hexagon in fig. 235 is formed by producing the sides of the regular hexagon. Prove that the area of the star-hexagon is twice that of the hexagon.

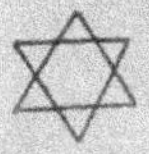

fig. 235.

Ex. **1193.** The area of the square circumscribed about a circle is twice the area of the square inscribed in the same circle.

Ex. **1194.** Prove that the area of the regular hexagon inscribed in a circle is twice the area of the inscribed equilateral triangle. Verify this fact by cutting a regular hexagon out of paper, and folding it.

Ex. **1195.** The side of an equilateral triangle circumscribed about a circle is twice the side of an inscribed equilateral triangle.

†Ex. **1196.** Chords AP, BQ are drawn perpendicular to a chord AB at its extremities. Prove that AP=BQ.

†Ex. **1197.** The line joining the centre of a circle to the point of intersection of two tangents is the perpendicular bisector of the line joining the points of contact of the tangents.

†Ex. **1198.** Find the locus of the point of intersection of tangents to a circle which meet at a constant angle.

Ex. **1199.** Show how to construct a right-angled triangle, given that the radius of the inscribed circle is 2 cm. and that one of the sides about the right angle is 5 cm.

Ex. **1200.** Show how to construct an isosceles triangle, given the radius of the inscribed circle, and the base.

†Ex. **1201.** A is a point outside a given circle (centre O, radius r). With centre O and radius $2r$ describe a circle; with centre A and radius AO describe a circle; let these two circles intersect at B, C. Let OB, OC cut the given circle at D, E. Prove that AD, AE are tangents to the given circle.

†Ex. **1202.** A circle is drawn having its centre on a side AC (produced) of an isosceles triangle, and touching the equal side AB at B. BC is produced to meet the circle at D. Prove that the radius of the circle through D is perpendicular to AC.

Ex. **1203.** Find the angles subtended at the centre of a circle by the three segments into which any tangent is divided by the sides (produced if necessary) of a circumscribed square.

†Ex. **1204.** An interior common tangent of two circles cuts the two exterior common tangents in A, B. Prove that AB is equal to the length intercepted on an exterior tangent between the points of contact.

†Ex. **1205**. The radius of the circumcircle of an equilateral triangle is twice the radius of the in-circle.

Ex. **1206**. Show how to inscribe three equal circles to touch one another in an equilateral triangle, of side 6 in. (fig. 236).

†Ex. **1207**. Two circles touch externally at E; AB, CD are parallel diameters drawn in the same sense; prove that AE, ED are in the same straight line; as also are BE, EC.

†Ex. **1208**. Two circles touch at A; T is any point on the tangent at A; from T are drawn tangents TP, TQ to the two circles. Prove that TP=TQ. What is the locus of points from which equal tangents can be drawn to two circles in contact?

fig. 236.

†Ex. **1209**. S is the circumcentre of a triangle ABC, and AD is an altitude. Prove that ∠BAD=∠CAS.

†Ex. **1210**. Through a given point on the circumference of a circle draw a chord which shall be bisected by a given chord. Give a proof.

Ex. **1211**. From the given angles, find all the angles of fig. 237.

†Ex. **1212**. Two circles intersect at B, C; P is a variable point on one of them. PB, PC (produced if necessary) meet the other circle at Q, R. Prove that QR is of constant length.

[Show that it subtends a constant angle at B.]

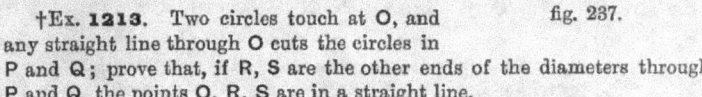

†Ex. **1213**. Two circles touch at O, and any straight line through O cuts the circles in

fig. 237.

P and Q; prove that, if R, S are the other ends of the diameters through P and Q, the points O, R, S are in a straight line.

†Ex. **1214**. A, B are the common points of two circles; C, D are points on the same arc AB of one of these circles. AC, AD meet the other circle again in G, H; BC, BD meet it again in E, F. Prove that the angles EAG, FBH are equal or supplementary.

Ex. **1215**. Show how to find a point O inside △ABC so that
$$\angle AOB = 150°, \quad \angle AOC = 130°.$$

Ex. **1216**. Show how to find a point O inside △ABC, such that the three sides subtend equal angles at O.

Ex. **1217.** Show how to construct a triangle, having given the vertical angle, the altitude and the bisector of the vertical angle (terminated by the base). In what case is this impossible?

†Ex. **1218.** A is one of the points of intersection of two circles whose centres are C, D. Through A is drawn a line PAQ, cutting the circles again in P, Q. PC, QD are produced to meet at R. Prove that the locus of R is a circle through C and D.

†Ex. **1219.** A, C are two fixed points, one upon each of two circles which intersect at B, D. Through B is drawn a variable chord PBQ, cutting the two circles in P, Q. PA, QC (produced if necessary) meet at R. Prove that the locus of R is a circle.

†Ex. **1220.** Two equal circles cut at A, B; a straight line PAQ meets the circles again in P, Q. Prove that BP=BQ. [Consider the angles subtended by the two chords.]

†Ex. **1221.** C is a variable point on a semicircle whose diameter is AB, centre O; CD is drawn ⊥ to AB; OX is the radius ⊥ to AB. On OC a point M is taken so that OM=CD. Prove that the locus of M is part of a circle whose diameter is OX.

†Ex. **1222.** ABC, DCB are two congruent triangles on the same side of the base BC. Prove that A, B, C, D are concyclic.

†Ex. **1223.** D, E, F are the mid-points of the sides BC, CA, AB of △ABC; AL is an altitude. Prove that D, E, F, L are concyclic (see Ex. 1222).

†Ex. **1224. Prove that the circle through the mid-points of the sides of a triangle also passes through the feet of the altitudes** (see Ex. 1223).

†Ex. **1225.** The altitudes BE, CF of △ABC intersect at H; prove
    (i) that AEHF is a cyclic quadrilateral,
    (ii) that ∠FAH=∠FEH,
    (iii) that ∠FEH=∠FCB,
    (iv) that, if AH is produced to meet BC in D, AFDC is cyclic,
    (v) that AD is ⊥ to CB.

Hence: **The three altitudes of a triangle meet in a point;** which is called the **orthocentre.**

†Ex. **1226.** In fig. 238 AD is ⊥ to BC and BE is ⊥ to CA; S is the centre of the circle. Show that

$$BF=AH,$$

and that AB, FH bisect one another.

[Prove AHBF a parallelogram.]

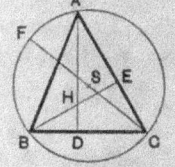

fig. 238.

†Ex. **1227.** BE, CF, two altitudes of △ABC, intersect at H. BE produced meets the circumcircle in K. Prove that E is the mid-point of HK.

[Show that BFEC is a cyclic quadrilateral, ∴ ∠FCE=∠FBE. But ∠KCE=∠FBE (why?), ∴ etc.]

†Ex. **1228.** I is the centre of the inscribed circle of △ABC; I_1 is the centre of the circle escribed outside BC. Prove that $BICI_1$ is cyclic.

†Ex. **1229.** An escribed circle of △ABC touches BC externally at D, and touches AB, AC produced at F, E respectively; O is the centre of the circle. Prove that

$$\text{(i)} \quad \angle BOC = \tfrac{1}{2}\angle FOE = 90° - \frac{A}{2},$$

$$\text{(ii)} \quad 2AE = 2AF = BC + CA + AB.$$

†Ex. **1230.** Prove that

$$\angle BIC = 90° + \frac{A}{2},$$

where I is the inscribed centre of △ABC.

Hence find the locus of the inscribed centre of a triangle, whose base and vertical angle are given.

†Ex. **1231.** I is the centre of the inscribed circle of △ABC; AI produced meets the circumcircle in P; prove that PB=PC=PI.

†Ex. **1232. P is any point on circumcircle of** △ABC. PL, PM, PN are ⊥ to BC, CA, AB re**spectively. Prove that**

    (i)   ∠PNL=180° − ∠PBC,

    (ii) ∠PNM=∠PAM,

    (iii) ∠PNL + ∠PNM=180°,

    (iv) LNM **is a straight line.**

Verify this result by drawing.

LNM is called Simson's line.

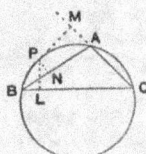

fig. 239.

†Ex. **1233.** ABCDEF is a regular hexagon; prove that BF is trisected by AC, AE.

†Ex. **1234.** In fig. 240, BC is ⊥ to PA. Prove that PA bisects ∠QPR.

†Ex. **1235.** Through A, a point of intersection of two circles, lines BAC, DAE are drawn, B, D being points on the one circle, C, E on the other. Prove that the angle between DB and CE (produced if necessary) is the same as the angle between the tangents at A.

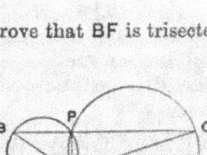

fig. 240.

†Ex. **1236.** Two circles touch internally at A; BC, a chord of the larger circle, touches the smaller at D; prove that AD bisects ∠BAC.

[Let BC meet the tangent at A in T.]

†Ex. **1237.** A radius of one circle is the diameter of another; prove that any straight line drawn from the point of contact to the outer circle is bisected by the inner circle.

†Ex. **1238.** In fig. 241 AB is a tangent;

$$OD = DA = AB.$$

BD cuts the circumference at E.   Prove that arc AE is $\frac{1}{12}$ and arc EF $\frac{1}{24}$ of the circumference.

†Ex. **1239.** Join O, the circumcentre of a triangle,   fig. 241. to the vertices A, B, C.   Through A draw lines ∥ to OB, OC; through B lines ∥ to OC, OA; through C lines ∥ to OA, OB.   Prove that these lines form an equilateral hexagon; that each angle of the hexagon is equal to one other angle, and double an angle of the triangle.

Ex. **1240.** Power is being transmitted from one shaft to another parallel shaft by means of a belt passing over two wheels.   The radii of the wheels are 2 ft. and 1 ft. and the distance between the shafts is 6 ft.   Assuming the belt to be taut, find its length (i) when it does not cross between the shafts, (ii) when it does cross.

†Ex. **1241.** PQ is a chord bisected by a diameter AB of a circle (centre O).   PG bisects the ∠OPQ.   Prove that it bisects the semicircle on which Q lies.

†Ex. **1242.** If through C, the mid-point of an arc AB, two chords are drawn, the first cutting the chord AB in D and the circle in E, the second cutting the chord in F and the circle in G, then the quadrilateral DFGE is cyclic.

†Ex. **1243.** P is a point on an arc AB.   Prove that the bisector of ∠APB and the perpendicular bisector of the chord AB meet on the circle.

†Ex. **1244.** P, Q are two points on a circle; AB is a diameter.   AP, AQ are produced to meet the tangents at B in X, Y.   Prove that △ˢ APQ, AYX are equiangular; and that P, Q, Y, X are concyclic.

†Ex. **1245.** In fig. 242 the angles at O are all equal; and OA=AB=BC=CD=DE. Prove that O, A, B, C, D, E are concyclic.

†Ex. **1246.** From a point A on a circle, two chords are drawn on opposite sides of the diameter through A. Prove that the line joining the mid-points of the minor arc of these chords cuts the chords at points equidistant from A.

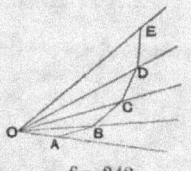

fig. 242.

†Ex. **1247.** Two equal chords of a circle intersect; prove that the segments of the one chord are respectively equal to the segments of the other.

Ex. **1248.** In fig. 243 O is the centre of the arc AB; and Q is the centre of the arc BC; ∠ADC is a right angle. DA=3, DC=5, DQ=3. Find OA and QC. (Let OD=x.)

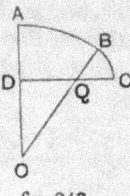

fig. 243.

†Ex. **1249.** CA, CB are two fixed radii of a circle. P is a variable point on the circumference; Q and R are the feet of the perpendiculars from P on CA and CB. Prove that QR is of constant length.

†Ex. **1250.** OADB is a rectangle; ACB is a triangle, right-angled at C, described on a diagonal AB of the rectangle. Prove that the angle between CD and OA is equal to the angle CAB.

†Ex. **1251.** Two circles X and Y touch at a point A; P and Q are points on X and Y respectively such that the tangents at P and Q meet on the tangent at A. Prove that a circle can be described through P and Q so as to touch both X and Y.

†Ex. **1252.** If one circle touch another internally at O, and any straight line cut the outer circle at A, D and the inner at B, C, prove that ∠AOB=∠COD.

†Ex. **1253.** T is any point outside a circle ABC whose centre is the point O. Through T two lines are drawn, TA touching the circle and TBC cutting it. If M is the mid-point of BC, show that ∠AMT=∠AOT.

If now AM be produced to cut the circle again in X, and XY be drawn parallel to BC to meet the circle again in Y, prove that TY is a tangent to the circle.

Ex. **1254.** On the same side of the straight line ADC two semicircles are drawn, touching at A; AKC has radius R and ALD has radius r. Find expressions for the radii of the following circles: (i) a circle which touches the two semicircles and has its centre equidistant from A and C, (ii) a circle which touches the two semicircles and has its centre equidistant from A and Ɔ.

Ex. **1255.** An egg-shaped channel, as shown in section in fig. 244, is formed by four circular arcs. The circles ABC and DEF touch in G, and the circular arcs AF and CD touch both circles, AC being a diameter of the larger circle. If the radii of ABC and DEF are 2 feet and 1 foot respectively, find the radius of the arcs AF and CD.

Find also the area enclosed by the figure.

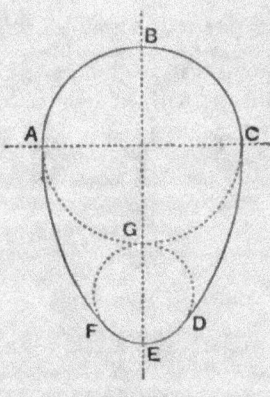

fig. 244.

Ex. **1256.** A and B are two fixed points on a circle. P is a point on the circle near to B. The circles on AB and AP as diameters meet at Q. What is the limiting position of Q when P coincides with B?

BOOK IV.

SIMILARITY.

Ratio and Proportion.

To **measure** a length is to find how many times it contains another length called the **unit** of length.

The unit of length may be an inch, a centimetre, a millimetre, a mile, a light-year*, or any length you choose. Hence the importance of always stating your unit.

If you have two lines, one 4 in. long, the other 5 in., you say that the first is $\frac{4}{5}$ of the second.

The **ratio** of a length XY to a length PQ is the quotient

$$\frac{\text{measure of XY}}{\text{measure of PQ}},$$

the two measurements being made with respect to the same unit of length.

The practical way then, to find the ratio of two lengths, is to measure them in inches or centimetres or any other convenient unit, and divide.

The ratio of a to b is written $\frac{a}{b}$, or a/b, or $a : b$, or $a \div b$.

---

* Astronomers sometimes express the distances of the fixed stars in terms of the distance traversed by light in a year. This distance is called a light-year, and is 63,368 times the distance of the earth from the sun. The nearest star is α Centauri, whose distance is 4·26 light-years.

¶Ex. **1257.** Find the ratio $\dfrac{PQ}{RS}$ (fig. 245); measure (i) in inches, (ii) in centimetres. Work out the ratios to three significant figures. Why might you expect your results to differ?

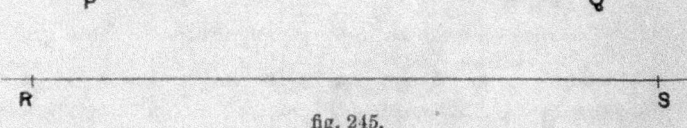

fig. 245.

Hitherto we have only considered the ratio of two lengths. In the case of other magnitudes, ratio may be defined as follows:

DEF. The **ratio** of one magnitude to another of the same kind is the quotient obtained by dividing the numerical measure of the first by that of the second, the unit being the same in each case.

The ratio of two magnitudes is independent of the unit chosen.

For example, the ratio of a length of 5 yds. to a length of 2 yds. is 5:2; if these lengths are measured in feet the measures are 15 and 6, and the ratio is 15:6. Now we know that $5:2=15:6$.

DEF. If $a:b=c:d$, the four magnitudes a, b, c, d are said to be **in proportion**.

¶Ex. **1258.** Fill in the missing terms in the following:

    (i) $\dfrac{2}{3}=\dfrac{}{6}$,   (iv) $5:2=7:$,

    (ii) $\dfrac{}{5}=\dfrac{1}{2}$,   (v) $\dfrac{}{2p}=\dfrac{3}{2}$,

    (iii) $7:\ =3:11$,   (vi) $\dfrac{a}{b}=\dfrac{}{d}$.

¶Ex. **1259.** Draw two straight lines SVT and XZY.

Prove that, if $\dfrac{SV}{ST}=\dfrac{XZ}{XY}$, then

    (i) $\dfrac{ST}{SV}=\dfrac{XY}{XZ}$,   (ii) $\dfrac{VT}{SV}=\dfrac{ZY}{XZ}$,   (iii) $\dfrac{VT}{ST}=\dfrac{ZY}{XY}$.

INTERNAL AND EXTERNAL DIVISION*.

If in a straight line AB a point P is taken, AB is said to be divided internally in the ratio $\frac{PA}{PB}$ (i.e. the ratio of the distances of P from the ends of the line). In the same way, if in AB produced a point P is taken, AB is said to be divided externally in the ratio $\frac{PA}{PB}$ (i.e. the ratio of the distances of P from the ends of the line).

In the latter case, it must be carefully noted that the ratio is not $\frac{AB}{BP}$. Suppose the points A, B connected by an elastic string; take hold of the string at a point P and, always keeping the three points in a straight line, vary the position of P; whether P is in AB or AB produced, the ratio in which AB is divided is always the ratio of the lengths of the two parts of the string.

¶Ex. **1260.** In fig. 246, name the ratios in which (i) H divides AB, (ii) A divides BH, (iii) C divides KA.

¶Ex. **1261.** In fig. 258, what lines are divided (i) by D in the ratio $\frac{BD}{DC}$, (ii) by Z in the ratio $\frac{ZY}{ZW}$, (iii) by B in the ratio $\frac{BC}{BD}$?

PROPORTIONAL DIVISION OF STRAIGHT LINES.

¶Ex. **1262.** Draw a triangle ABC and draw HK parallel to BC (see fig. 246). What fraction is AH of AB? What fraction is AK of AC?

[Express these fractions as decimals.]

* The discussion of cases of external division may be postponed.

Theorem 1.

If a straight line HK drawn parallel to the base BC of a triangle ABC cuts AB, AC in H, K respectively, then $\dfrac{AH}{AB} = \dfrac{AK}{AC}$.

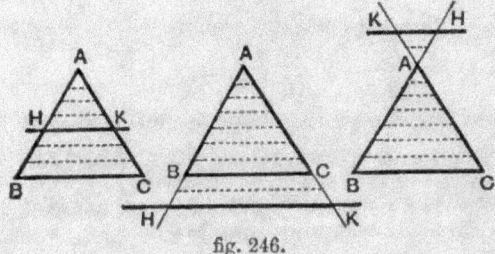

fig. 246.

Proof Suppose that $\dfrac{AH}{AB} = \dfrac{p}{q}$, where p and q are integers. Then if AB is divided into q equal parts, AH contains p of these parts.

Through the points of division draw parallels to BC.

Now AB is divided into equal parts.

∴ these parallels divide AC into equal parts;   I. 24.

AC contains q of these parts, and AK contains p of these parts.

$$\therefore \frac{AK}{AC} = \frac{p}{q},$$

$$\therefore \frac{AH}{AB} = \frac{AK}{AC}.$$

Q. E. D.

Cor. 1. **If a straight line is drawn parallel to one side of a triangle, the other two sides are divided proportionally.**

To prove that $\dfrac{AH}{HB} = \dfrac{AK}{KC}$.

*First proof** In the figure, AB is divided into q equal parts;

AH contains p of these equal parts;

$\therefore$ HB   ,, $q - p$,,   ,,

$\therefore \dfrac{AH}{HB} = \dfrac{p}{q - p}$.

$\text{Sim}^{ly} \dfrac{AK}{KC} = \dfrac{p}{q - p}$.

$\therefore \dfrac{AH}{HB} = \dfrac{AK}{KC}$.

Q. E. D.

*Second proof**   Since $\dfrac{AH}{AB} = \dfrac{AK}{AC}$,   *Proved*

$\therefore \dfrac{AB}{AH} = \dfrac{AC}{AK}$,

$\therefore \dfrac{AB}{AH} - 1 = \dfrac{AC}{AK} - 1$,

$\therefore \dfrac{AB - AH}{AH} = \dfrac{AC - AK}{AK}$,

i.e. $\dfrac{HB}{AH} = \dfrac{KC}{AK}$,

$\therefore \dfrac{AH}{HB} = \dfrac{AK}{KC}$.

Q. E. D.

* These proofs apply to the first figure: see Ex. 1263.

Cor. 2. **If two straight lines are cut by a series of parallel straight lines, the intercepts on the one have to one another the same ratios as the corresponding intercepts on the other.**

†Ex. **1263**. Write out the two proofs of Cor. 1 for the second and third figures of page 222.

†Ex. **1264**. Triangles of the same height are to one another as their bases.

[Suppose one base is $\frac{2}{3}$ of the other.]

Ex. **1265**. **Show how to divide a given straight line AB in the ratio of two given straight lines** p, q.

[Through A draw AC, from AC cut off AD$=p$, DE$=q$; join BE; draw a line through D to divide AB in the ratio $\dfrac{AD}{DE}$; in what direction must this line be drawn?]

Ex. **1266**. Find the value of x, when $\dfrac{2\cdot5}{4\cdot2}=\dfrac{x}{3\cdot7}$, (i) graphically, (ii) by calculation.

[Make an ∠POQ; from OP cut off OD$=4\cdot2$ in., OE$=2\cdot5$ in.; from OQ cut off OF$=3\cdot7$ in.; draw EG ∥ to DF. Which is the required length?]

Ex. **1267**. Find, both graphically and by calculation, the value of x in the following cases :

(i) $\dfrac{2\cdot25}{3\cdot05}=\dfrac{3\cdot05}{x}$,   (ii) $\dfrac{\cdot935}{x}=\dfrac{1\cdot225}{5\cdot75}$,

(iii) $x:2\cdot63=5\cdot05:2\cdot84$,   (iv) $8\cdot36:\cdot025=x:\cdot037$.

Def. If x is such a magnitude that $\dfrac{a}{b}=\dfrac{c}{x}$ (or $a:b=c:x$), x is called the **fourth proportional** to the three magnitudes a, b, c.

Def. If x is such a magnitude that $\dfrac{a}{b}=\dfrac{b}{x}$ (or $a:b=b:x$), x is called the **third proportional** to the two magnitudes a, b.

Note. If x is the third proportional to a, b, it is also the fourth proportional to a, b, b.

To find the fourth proportional to three given straight lines.

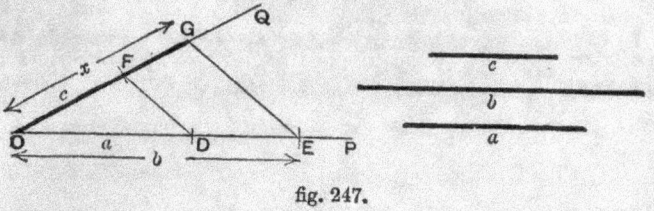

fig. 247.

Let a, b, c be the three given straight lines.

Construction   Make an angle **POQ**.

From **OP** cut off **OD** = a, and **OE** = b.

From **OQ** cut off **OF** = c.

Join **DF**.

Through **E** draw **EG** ∥ to **DF**, cutting **OQ** in **G**.

Then **OG** (x) is the fourth proportional to a, b, c.

Proof   Since **FD** is ∥ to **EG**,

$$\therefore \frac{a}{b} = \frac{c}{x}.$$ IV. 1.

Ex. **1268.** **Show how to find the third proportional to two given straight lines.**

[See note above.]

Ex. **1269.** Find graphically the fourth proportional to 3, 4, 5. Check by calculation.

Ex. **1270.** Find graphically the third proportional to 6·32, 8·95. Check by calculation.

Ex. **1271.** Draw a straight line AB, on it take two points P, Q; draw another straight line CD; show how to divide CD similarly to AB.

THEOREM 2.

[CONVERSE OF THEOREM 1.]

If H, K are points in the sides AB, AC of a triangle ABC, such that $\dfrac{AH}{AB} = \dfrac{AK}{AC}$, then HK is parallel to BC.

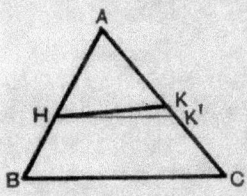

fig. 248.

Construction	Draw HK′ parallel to BC.
To prove that	HK and HK′ coincide.
Proof	Since HK′ is ∥ to BC,

$$\therefore \frac{AH}{AB} = \frac{AK'}{AC}. \qquad\qquad \text{IV. 1.}$$

$$\text{But } \frac{AH}{AB} = \frac{AK}{AC}. \qquad\qquad Data$$

$$\therefore \frac{AK'}{AC} = \frac{AK}{AC},$$

$$\therefore AK' = AK,$$

$\therefore$ K and K′ coincide,

$\therefore$ HK and HK′ coincide,

$\therefore$ HK is parallel to BC.

Q. E. D.

COR. 1.   If $\dfrac{AB}{AH} = \dfrac{AC}{AK}$, then HK and BC are parallel.

COR. 2.   If a straight line divides the sides of a triangle proportionally, it is parallel to the base of the triangle.

†Ex. **1272.** Prove Cor. 1 without assuming IV. 2.

†Ex. **1273.** Prove Cor. 2 without assuming IV. 2.

†Ex. **1274.** O is a point inside a quadrilateral ABCD; OA, OB, OC, OD are divided at A′, B′, C′, D′

so that
$$\frac{OA'}{OA} = \frac{OB'}{OB} = \frac{OC'}{OC} = \frac{OD'}{OD} = \frac{2}{3}.$$

Prove that A′B′ is parallel to AB.

Also prove that ∠D′A′B′ = ∠DAB.

†Ex. **1275.** Draw a triangle ABC; in it take a point O, and join OA, OB, OC; in OA take a point A′, through A′ draw A′B′ parallel to AB to cut OB at B′, through B′ draw B′C′ parallel to BC to cut OC at C′. Prove that C′A′, CA are parallel.

†Ex. **1276.** A variable line, drawn through a fixed point O, cuts two fixed parallel straight lines at P, Q; prove that OP : OQ is constant.

†Ex. **1277.** O is a fixed point and P moves along a fixed line. OP is divided at Q (internally or externally) in a fixed ratio. Find the locus of Q.

†Ex. **1278.** D is a point in the side AB of △ABC; DE is drawn parallel to BC and cuts AC at E; EF is drawn parallel to AB and cuts BC at F. Prove that AD : DB = BF : FC.

†Ex. **1279.** D is a point in the side AB of △ABC; DE is drawn parallel to BC and cuts AC at E; CF is drawn parallel to EB and cuts AB produced at F. Prove that AD : AB = AB : AF.

†Ex. **1280.** AD, BC are the parallel sides of a trapezium; prove that a line drawn parallel to these sides cuts the other sides proportionally.

†Ex. **1281.** From a point E in the common base AB of two triangles ACB, ADB, straight lines are drawn parallel to AC, AD, meeting BC, BD at F, G; show that FG is parallel to CD.

†Ex. **1282.** In three straight lines OAP, OBQ, OCR the points are chosen so that AB is parallel to PQ, and BC parallel to QR. Prove that AC is parallel to PR.

†Ex. **1283.** AB, DC are the parallel sides of a trapezium. P, Q are points on AD, BC, so that AP/PD = BQ/QC. Prove that PQ is ∥ to AB and DC. (Use *reductio ad absurdum*.)

15—2

SIMILAR FIGURES.

Figures of the same shape are said to be **similar**. For instance, any two squares are similar: any two circles are similar. A picture is similar to a reduced photograph of it; a picture thrown on a screen by a magic lantern is similar to the picture on the lantern slide.

If we have a figure of a polygon on a lantern slide, and compare it with the similar polygon on the screen, we see (i) that the angles are unaltered, (ii) that the sides are all magnified in the same proportion. This suggests the following definition.

DEF. Polygons which are equiangular to one another and have their corresponding sides proportional are called **similar** polygons.

¶Ex. **1284.** Is a square equiangular with a rectangle? Is it similar?

¶Ex. **1285.** Have a square and a rhombus their corresponding sides proportional? Are they similar?

¶Ex. **1286.** A rectangular picture frame is made of wood 1 in. wide; are the inside and outside of the frame similar rectangles?

¶Ex. **1287.** Draw a quadrilateral ABCD; draw a straight line parallel to CD to cut BC at P and AD at Q. Prove that ABCD, ABPQ are equiangular. Are they similar?

¶Ex. **1288.** Are two equiangular triangles similar?

THEOREM 3.

If two triangles are equiangular, their corresponding sides are proportional.

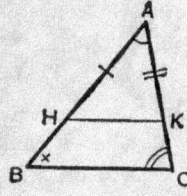 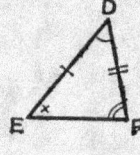

fig. 249.

Data　　　ABC, DEF are two triangles which have

∠A = ∠D, ∠B = ∠E, and ∠C = ∠F.　(See I. 8, Cor. 5.)

To prove that　　　$\dfrac{BC}{EF} = \dfrac{CA}{FD} = \dfrac{AB}{DE}$.

Construction　　　From AB cut off AH = DE,

From AC cut off AK = DF.

Join HK.

Proof　　　In the △s AHK, DEF,

　　　$\begin{cases} AH = DE, & \textit{Constr.} \\ AK = DF, & \textit{Constr.} \\ ∠A = ∠D, & \textit{Data} \end{cases}$

∴ △AHK ≡ △DEF,　　　I. 10.

∴ ∠AHK = ∠E,

= ∠B.　　　*Data*

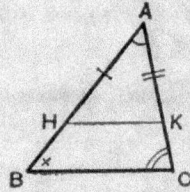
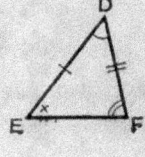

fig. 250.

$$\therefore \text{ HK is } \| \text{ to BC,} \qquad\qquad\qquad \text{I. 4.}$$

$$\therefore \frac{AH}{AB} = \frac{AK}{AC}, \qquad\qquad\qquad \text{IV. 1.}$$

$$\therefore \frac{DE}{AB} = \frac{DF}{AC}.$$

Sim^{ly} by cutting off lengths from BA, BC, $\dfrac{ED}{BA} = \dfrac{EF}{BC}$

$$\therefore \frac{EF}{BC}, \frac{FD}{CA}, \frac{DE}{AB} \text{ are all equal.}$$

$$\therefore \frac{BC}{EF} = \frac{CA}{FD} = \frac{AB}{DE}.$$

Q. E. D.

†Ex. **1289.** Write out the complete proof that $\dfrac{ED}{BA} = \dfrac{EF}{BC}$.

Ex. **1290.** ABC is a triangle having BC=3 in., CA=4 in., AB=5 in.; DEF is an equiangular triangle having EF=2·2 in. Calculate DE, DF.

Ex. **1291.** Repeat Ex. 1290 with BC=5·8 cm., CA=7·7 cm., AB=8·3 cm., EF=1·8 in.

¶Ex. **1292.** If P is any point on either arm of an angle XOY, and PN is drawn perpendicular to the other arm, $\dfrac{PN}{OP}$ has the same value for all positions of P.

[Take several different positions of P and prove that $\dfrac{PN}{OP} = \dfrac{P_1N_1}{OP_1} = \ldots$]

$\dfrac{PN}{OP}$ is the **sine** of $\angle$XOY; this exercise might have been stated as follows:—the sine of an angle depends only on the magnitude of the angle.

¶Ex. **1293.** Prove that the **cosine** $\left(\dfrac{ON}{OP}\right)$ and **tangent** $\left(\dfrac{PN}{ON}\right)$ of an angle depend only on the magnitude of the angle.

†Ex. **1294.** PQRS is a quadrilateral inscribed in a circle whose diagonals intersect at X; prove that the △ˢ XPS, XQR are equiangular. Write down the three equal ratios of corresponding sides.

†Ex. **1295.** In the figure of Ex. 1294, prove that $\dfrac{PQ}{SR}=\dfrac{XP}{XS}$.

[If you colour PQ, SR red, and XP, XS blue, you will see which two triangles you require.]

†Ex. **1296.** XYZW is a cyclic quadrilateral; XY, WZ produced intersect at a point P outside the circle; prove that $\dfrac{PY}{PW}=\dfrac{PZ}{PX}$.

†Ex. **1297.** ABC is a triangle right-angled at A; prove that the altitude AD divides the triangle into two triangles which are similar to △ ABC. Write down the ratio properties you obtain from the similarity of △ˢ BDA, BAC.

†Ex. **1298.** The altitude QN of a triangle PQR right-angled at Q cuts RP in N; prove that $\dfrac{QN}{RN}=\dfrac{PN}{QN}$.

[Find two equiangular triangles; colour the given lines; see Ex. 1295.]

†Ex. **1299.** XYZ is a triangle inscribed in a circle, XN is an altitude of the triangle, and XD a diameter of the circle; prove that

$$XY:XD=XN:XZ.$$

†Ex. **1300.** XYZ is a triangle inscribed in a circle; the bisector of ∠X meets YZ in P, and the circle in Q; prove that XY : XQ = XP : XZ.

†Ex. **1301.** PQRS is a quadrilateral inscribed in a circle; PT is drawn so that ∠SPT = ∠QPR. (See fig. 251.) Prove that (i) SP : PR = ST : QR,
(ii) SP : PT = SR : TQ.

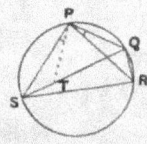

†Ex. **1302.** Three straight lines are drawn from a point O; they are cut by a pair of parallel lines at X, Y, Z and X′, Y′, Z′. Prove that XY : YZ = X′Y′ : Y′Z′.

fig. 251.

Ex. **1303.** On a base 4 in. long draw a quadrilateral; on a base 3 in. long construct a similar quadrilateral. Calculate the ratio of each pair of corresponding sides. [Draw a diagonal of the first quadrilateral.]

On a given straight line to construct a figure similar to a given rectilinear figure.

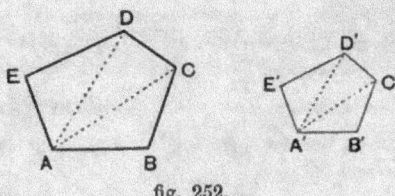

fig. 252.

Let ABCDE be the given figure and A′B′ the given straight line.

Construction   Join AC, AD.
On A′B′ make △A′B′C′ equiangular to △ABC.
On A′C′ make △A′C′D′ equiangular to △ACD.
On A′D′ make △A′D′E′ equiangular to △ADE.
Then A′B′C′D′E′ is similar to ABCDE.

Proof   This may be divided into two parts:

(i) the proof that the figures are equiangular; this is left to the student.

(ii) the proof that $\dfrac{AB}{A'B'} = \dfrac{BC}{B'C'} = \dfrac{CD}{C'D'} = \dfrac{DE}{D'E'} = \dfrac{EA}{E'A'}$.

Since △ˢ ABC, A′B′C′ are equiangular,   *Constr.*

∴ $\dfrac{AB}{A'B'} = \dfrac{BC}{B'C'} = \dfrac{AC}{A'C'}$.   IV. 3.

Since △ˢ ACD, A′C′D′ are equiangular,   *Constr.*

∴ $\dfrac{AC}{A'C'} = \dfrac{CD}{C'D'} = \dfrac{AD}{A'D'}$.   IV. 3.

Again since △ˢ ADE, A′D′E′ are equiangular,   *Constr.*

∴ $\dfrac{AD}{A'D'} = \dfrac{DE}{D'E'} = \dfrac{EA}{E'A'}$,   IV. 3.

∴ $\dfrac{AB}{A'B'} = \dfrac{BC}{B'C'} = \dfrac{CD}{C'D'} = \dfrac{DE}{D'E'} = \dfrac{EA}{E'A'}$.

For exercises, see p. 234.

THEOREM 4.

[CONVERSE OF THEOREM 3.]

If, in two triangles ABC, DEF, $\dfrac{BC}{EF} = \dfrac{CA}{FD} = \dfrac{AB}{DE}$, then the triangles are equiangular.

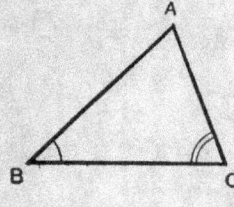 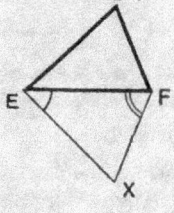

fig. 253.

Construction   Make $\angle$ FEX = $\angle$ B and $\angle$ EFX = $\angle$ C, X and D being on opposite sides of EF.

Proof   In the $\triangle$s ABC, XEF,

$\begin{cases} \angle\,B = \angle\,FEX, \\ \angle\,C = \angle\,EFX, \end{cases}$

∴ the third angles are equal,
and the triangles are equiangular.

∴ $\dfrac{BC}{EF} = \dfrac{CA}{FX} = \dfrac{AB}{XE}.$ IV. 3.

But $\dfrac{BC}{EF} = \dfrac{CA}{FD} = \dfrac{AB}{DE}$,   *Data*

∴ $\dfrac{CA}{FX} = \dfrac{CA}{FD}$ and $\dfrac{AB}{XE} = \dfrac{AB}{DE}$,

∴ FX = FD and XE = DE.

In the $\triangle$s XEF, DEF,

$\begin{cases} XE = DE, \\ FX = FD, \\ \text{and EF is common,} \end{cases}$

∴ $\triangle$ XEF ≡ $\triangle$ DEF.   I. 14.

But the $\triangle$s ABC, XEF are equiangular,
∴ the $\triangle$s ABC, DEF are equiangular.   Q. E. D.

†Ex. **1304**. Prove that two quadrilaterals are similar if their sides and one of their diagonals are proportional.

The **diagonal scale** (fig. 254) depends in principle on the properties of similar triangles.

¶Ex. **1305**. 'Are the triangles whose corners are marked 0, *d*, 10 and 0, *c*, 6 equiangular?

¶Ex. **1306**. What fraction is the distance between the points 6, *c* of the distance between 10, *d*?

The distance between 10, *d* is ·1 in.; what is the distance between 6, *c*?

¶Ex. **1307**. What are the distances between the points (i) *a*, 6, (ii) 6, *c*, (iii) *c*, *b*?

What is the whole distance between *a*, *b*?

Ex. **1308**. On inch-paper, mark the points O (0, 0), P (3, 0), Q (5, 2), R (4, 5), S (1, 4); join OP, PQ, QR, RS, SO. On plain paper, draw O'P = 1·5 in.; on O'P' describe a similar polygon.

Ex. **1309**. Draw a pentagon ABCDE; draw A'B' ∥ to AB; on A'B' construct a pentagon similar to ABCDE. (This should be done with set-square and straight edge only.)

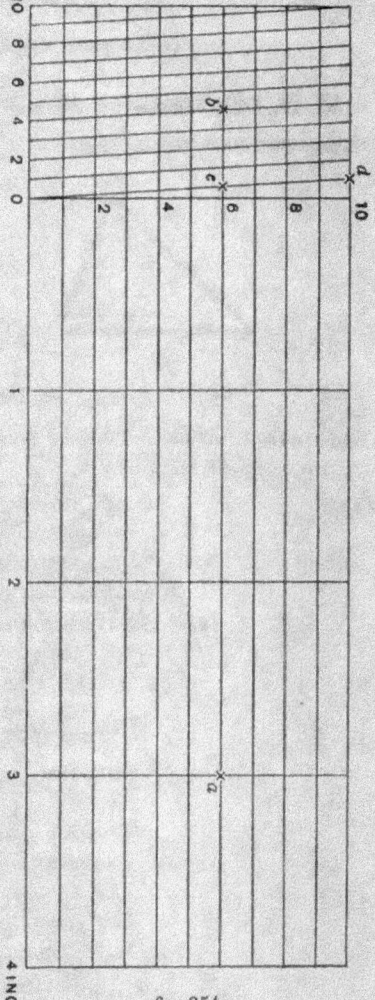

fig. 254.

THEOREM 5.

If two triangles have one angle of the one equal to one angle of the other and the sides about these equal angles proportional, the triangles are similar.

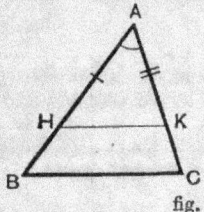 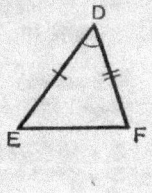

fig. 255.

Data   ABC, DEF are two triangles which have ∠ A = ∠ D, and

$$\frac{AB}{DE} = \frac{AC}{DF}.$$

To prove that   the △s ABC, DEF are similar.

Construction   From AB cut off AH = DE.

From AC cut off AK = DF.

Join HK.

Proof   In △s AHK, DEF,

AH = DE,	*Constr.*
AK = DF,	*Constr.*
∠ A = ∠ D,	*Data*

∴ △ AHK ≡ △ DEF.   I. 10.

Now $\dfrac{AB}{DE} = \dfrac{AC}{DF}.$

∴ $\dfrac{AB}{AH} = \dfrac{AC}{AK},$

∴ HK is ‖ to BC.   IV. 2, *Cor.* 1.

∠ H = ∠ B and ∠ K = ∠ C,

∴ △s AHK, BCA are equiangular.

Hence △s DEF, ABC are equiangular,

and therefore have their corresponding sides

proportional.   IV. 3.

∴ △s DEF, ABC are similar.   Q. E. D.

NOTE.   In IV. 3 and 5, if DE > AB and DF > AC, H, K lie in AB, AC produced ; the proofs hold equally well for these cases.

†Ex. **1310.** S is a point in the side PQ of △ PQR ; ST is drawn parallel to QR and of such a length that ST : QR = PS : PQ. Prove that T lies in PR.

[Prove ∠ SPT = ∠ QPR.]

Ex. **1311.** (Inch-paper.) Prove that the points (0, 0), (2, 1), (5, 2·5) are in a straight line. In what ratio is the line divided?

†Ex. **1312.** In a triangle ABC, AD is drawn perpendicular to the base ; if BD : DA = DA : DC, prove that △ ABC is right-angled.

†Ex. **1313.** AX, DY are medians of the two similar triangles ABC, DEF ; prove that they make equal angles with BC, EF, and that AX : DY = AB : DE. (Compare Ex. 283.)

†Ex. **1314.** The bases, BC, EF, of two similar triangles, ABC, DEF, are divided in the same ratio at X, Y. Prove that AX : DY = BC : EF.

Fig. 256 represents a pair of **proportional compasses.** AB = AC and AH = AK,

$$\therefore \frac{AH}{AB} = \frac{AK}{AC},\ \text{and}\ \angle\ BAC = \angle\ HAK,$$

$$\therefore\ \triangle^s\ ABC,\ AHK\ \text{are similar.}$$

Hence $\frac{HK}{BC} = \frac{AH}{AB}$, which is constant for any fixed position of the hinge. In fig. 256 the hinge is adjusted so that $\frac{AH}{AB} = \frac{1}{2}$; thus, whatever the angle to which the compasses are opened, HK = $\frac{1}{2}$ BC.

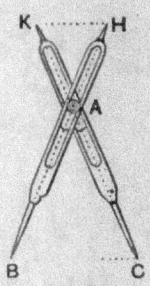

fig. 256.

AREAS OF SIMILAR FIGURES.

Consider any figure, rectilinear or curvilinear; let it be divided up into small equal squares for purposes of counting the number of units of area. Now suppose that the figure, squares and all, is magnified, say 3 times; we shall have a figure similar

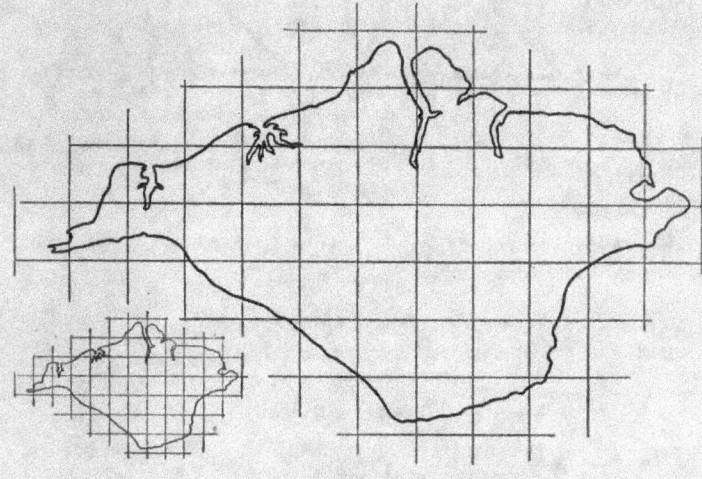

fig. 257.

to the former, the linear dimensions being in the ratio $1:3$. Each of the small squares will have its area increased 9 times; and as the number of squares is unchanged, the area of the whole figure will be magnified 9 times.

This suggests that **the ratio of the areas of similar figures is equal to the square of the ratio of their linear dimensions.** A formal proof follows for the case of the triangle.

†Ex. **1315.** The ratio of corresponding altitudes of similar triangles is equal to the ratio of corresponding sides.

THEOREM 6.

The ratio of the areas of similar triangles is equal to the ratio of the squares on corresponding sides.

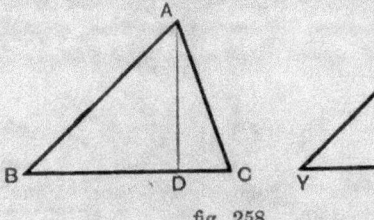

fig. 258.

Data   ABC, XYZ are two similar triangles,

To prove that $\dfrac{\triangle ABC}{\triangle XYZ} = \dfrac{BC^2}{YZ^2}$.

Construction   Draw AD ⊥ to BC,
          and XW ⊥ to YZ.

Proof $\triangle ABC = \frac{1}{2}BC \cdot AD$,   II. 2.
          and $\triangle XYZ = \frac{1}{2}YZ \cdot XW$,
          $\therefore \dfrac{\triangle ABC}{\triangle XYZ} = \dfrac{BC \cdot AD}{YZ \cdot XW}$.

$\left[\text{It remains to prove that } \dfrac{AD}{XW} = \dfrac{BC}{YZ}.\right]$

Now in the △s ABD, XYW,

$\begin{cases} \angle B = \angle Y, \\ \angle D = \angle W \text{ (rt. } \angle s), \end{cases}$ *Data*

∴ the third angles are equal,
and the △s are equiangular,

$\therefore \dfrac{AD}{XW} = \dfrac{AB}{XY}$.   IV. 3.

But $\dfrac{AB}{XY} = \dfrac{BC}{YZ}$,   *Data*

$\therefore \dfrac{AD}{XW} = \dfrac{BC}{YZ}$.

$$\text{But } \frac{\triangle ABC}{\triangle XYZ} = \frac{BC}{YZ} \cdot \frac{AD}{XW} \qquad\qquad Proved$$

$$= \frac{BC}{YZ} \cdot \frac{BC}{YZ}$$

$$= \frac{BC^2}{YZ^2}.$$

Q. E. D.

¶Ex. **1316.** What is the ratio of the areas of two similar triangles on bases of 3 in. and 4 in.?

¶Ex. **1317.** The area of a triangle with a base of 12 cm. is 60 sq. cm.; find the area of a similar triangle with a base of 9 cm.

What is the area of a similar triangle on a base of 9 in.?

¶Ex. **1318.** The areas of two similar triangles are 100 sq. cm. and 64 sq. cm.; the base of the greater is 7 cm.; find the base of the smaller.

¶Ex. **1319.** What is the ratio of the area of a room to the area by which it is represented on a plan whose scale is 1 in. to 1 ft.?

¶Ex. **1320.** On a map whose scale is 1 mile to 1 in., a piece of land is represented by an area of 20 sq. in.; what is the area of the land?

¶Ex. **1321.** On a map whose scale is 2 miles to 1 in., a piece of land is represented by an area of 24 sq. in.; what is the area of the land?

Ex. **1322.** What is the acreage of a field which is represented by an area of 3 sq. in. on a map whose scale is 25 in. to the mile? (640 acres = 1 sq. mile.)

Ex. **1323.** Two similar windows are glazed with small lozenge-shaped panes of glass, these panes being all identical in size and shape. The heights of the windows are 10 ft. and 15 ft. The number of panes in the smaller window is 1200; what is the number in the larger?

Ex. **1324.** What is the ratio of the areas of two circles whose radii are (i) R, r, (ii) 3 in., 2 in.?

Ex. **1325.** The load that can be carried by an aeroplane travelling at a given velocity depends on the area of its wings. Experiments with a model show that it can carry a weight w lbs.; the full-sized machine is 10 times the length of the model; what weight can the machine carry at the same velocity?

Ex. **1326.** Similar figures are described on the side and diagonal of a square; prove that the ratio of their areas is 1 : 2.

Ex. **1327.** Similar figures are described on the side and altitude of an equilateral triangle; prove that the ratio of their areas is 4 : 3.

Ex. **1328.** Show how to draw a straight line parallel to the base of a triangle to bisect the triangle.

†Ex. **1329.** Prove that, if similar triangles are described on the three sides of a right-angled triangle. the area of the triangle described on the hypotenuse is equal to the sum of the other two triangles.

Ex. **1330.** A figure described on the hypotenuse of a right-angled triangle is equal to the sum of the similar figures described on the sides of the triangle. (This is a generalisation of Pythagoras' theorem.)

†Ex. **1331.** ABC, DEF are two triangles in which $\angle B = \angle E$; prove that $\triangle ABC : \triangle DEF = AB \cdot BC : DE \cdot EF$.

[Draw AX $\perp$ to BC, and DY $\perp$ to EF.]

RECTANGLE PROPERTIES.

†Ex. **1332.** XYZ is a triangle inscribed in a circle, XN is an altitude of the triangle and XD a diameter of the circle; prove that $\dfrac{XY}{XN} = \dfrac{YD}{NZ}$. Express this as a result clear of fractions. What two rectangles are thus proved equal?

†Ex. **1333.** With the same construction as in Ex. 1332, prove that

$$XZ \cdot NY = XN \cdot ZD.$$

[You will have to pick out two equal ratios from two equiangular triangles. If you colour XZ, NY red and XN, ZD blue you will see which are the triangles.]

†Ex. **1334.** ABCD is a quadrilateral inscribed in a circle; its diagonals intersect at X. Prove that (i) AX $\cdot$ BC = AD $\cdot$ BX, (ii) AX $\cdot$ XC = BX $\cdot$ XD.

†Ex. **1335.** ABCD is a quadrilateral inscribed in a circle; AB, DC produced intersect at Y. Prove that

(i) YA . BD=YD . CA, (ii) YA . YB=YC . YD.

†Ex. **1336.** The rectangle contained by two sides of a triangle is equal to the rectangle contained by the diameter of the circumcircle and the altitude drawn to the base.

[Draw the diameter through the vertex at which the two sides intersect.]

†Ex. **1337.** The bisector of the angle A of △ ABC meets the base in P and the circumcircle in Q. Prove that the rectangle contained by the sides AB, AC=rect. AP . AQ.

†Ex. **1338.** In Ex. 1301, prove that PQ . SR=PR . TQ.

†Ex. **1339. The sum of the rectangles contained by opposite sides of a cyclic quadrilateral is equal to the rectangle contained by its diagonals. (Ptolemy's theorem.)** [Use the construction of Ex. 1301.]

¶Ex. **1340.** Draw a circle and a diameter AOB; in OB mark a point P. Now rotate AB about P; note that at first PB increases and PA decreases. What will happen to the product PA . PB? Take two positions of the chord and compare their product.

¶Ex. **1341.** Repeat Ex. 1340 for the case in which P is in AB produced.

THEOREM 7 (i).

If AB, CD, two chords of a circle, intersect at a point P inside the circle, then PA . PB = PC . PD.

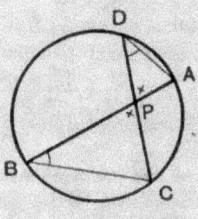

fig. 259.

Construction   Join BC, AD.

Proof   In the △s PAD, PCB,

$$\begin{cases} \angle APD = \angle CPB \text{ (vert. opp.),} \\ \quad \angle B = \angle D \text{ (in the same segment),} \end{cases}$$

∴ the third angles are equal,

and the △s are equiangular,

$$\therefore \frac{PA}{PC} = \frac{PD}{PB},$$ IV. 3.

∴ PA . PB = PC . PD.

Q. E. D.

To calculate the area of the rectangle PA . PB in IV. 7 (i).

Suppose EPF is the chord bisected at P.

Then PA . PB = PE . PF = PE^2 = OE^2 − OP^2.

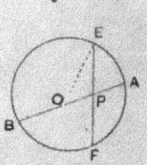

fig. 260.

THEOREM 7 (ii).

If AB, CD, two chords of a circle, intersect at a point P outside the circle, then PA . PB = PC . PD.

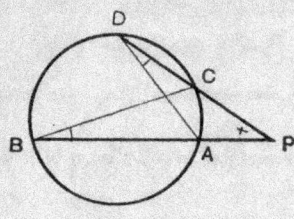

fig. 261.

Construction   Join BC, AD.

Proof   In the △s PAD, PCB,

$\begin{cases} \angle\text{ P is common,} \\ \angle\text{ B} = \angle\text{ D (in the same segment),} \end{cases}$

∴ the third angles are equal,

and the △s are equiangular,

$$\therefore \frac{PA}{PC} = \frac{PD}{PB},$$ IV. 3.

∴ PA . PB = PC . PD.

Q. E. D.

Cor. If PT is a tangent to a circle and AB a chord of the circle passing through P, then PT² = PA . PB. (See fig. 262.)

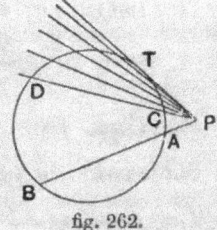

fig. 262.

16—2

Note. Theorems 7 (i) and 7 (ii) are really two different cases of the same theorem; notice that the proofs are nearly identical.

To calculate the area of the rectangle PA.PB in IV. 7 (ii).

See fig. 262. Use the fact that

$$PA . PB = PT^2 = OP^2 - OT^2.$$

†Ex. **1342.** Prove the corollary (i) by the method of limits, (ii) by similar triangles.

¶Ex. **1343.** What becomes of IV. 7 when P is a point *on* the circle? When P is the centre?

Ex. **1344.** Calculate the areas of the rectangles contained by the segments of chords passing through P when (i) $r = 5$ in., $OP = 3$ in., (ii) $r = 5$ cm., $OP = 13$ cm.

Ex. **1345.** Find an expression for the areas in Ex. 1344, r being the radius, and d the distance OP (i) when $d < r$, (ii) when $d > r$.

Ex. **1346.** Draw two straight lines APB, CPD intersecting at P; make PA = 4 cm., PB = 6 cm., PC = 3 cm. Describe a circle through ABC, cutting CP produced in D. Calculate PD. (*Freehand.*)

What would be the result if the exercise were repeated with the same lengths, but a different angle between APB, CPD?

Ex. **1347.** From a point P draw two straight lines PAB, PC; make PA = 4 cm., PB = 9 cm., PC = 6 cm. Describe a circle through ABC; let it cut PC again at D. Calculate PD. (*Freehand.*)

†Ex. **1348.** APB, CPD intersect at P; and the lengths PA, PB, PC, PD are so chosen that PA.PB = PC.PD. Prove that A, B, C, D are concyclic. (Draw ⊙ through ABC; let it cut CP produced in D'.) What relation does this exercise bear to IV. 7 (i)?

†Ex. **1349.** State and prove the converse of IV. 7 (ii).

†Ex. **1350.** P is a point outside a circle ABC and straight lines PAB, PC are drawn (A, B, C being on the circle); prove that, if $PA.PB = PC^2$, PC is the tangent at C. [Use *reductio ad absurdum.*]

†Ex. **1351**. ABC is a triangle right-angled at A; AD is drawn perpendicular to BC; prove that $AD^2 = BD \cdot DC$.

[Produce AD to cut the circumcircle of △ABC.]

†Ex. **1352**. If the common chord of two intersecting circles be produced to any point T, the tangents to the circles from T are equal to one another.

†Ex. **1353**. The common chord of two intersecting circles bisects their common tangents.

†Ex. **1354**. The altitudes BE, CF of a triangle ABC intersect at H, prove that

(i) BH . HE=CH . HF,   (ii) AF . AB=AE . AC,   (iii) BH . BE=BF . BA.

†Ex. **1355**. Two circles intersect at A, B; T is any point in AB, or AB produced; TCD, TEF are drawn cutting the one circle in C, D, the other in E, F. Prove that C, D, E, F are concyclic.

———

†Ex. **1356**. ABC is a triangle right-angled at A; AD is an altitude of the triangle. Prove that △^{s} ABD, CDA are equiangular. Write down the three equal ratios.

Def. If x is such a quantity that $a : x = x : b$, then x is called the **mean proportional** between a and b.

Note. If $\dfrac{a}{x} = \dfrac{x}{b}$, $x^2 = ab$, and therefore $x = \sqrt{ab}$; thus the mean proportional between two numbers is the square root of the product.

¶Ex. **1357**. Find the mean proportional between

(i) 4 and 9,   (ii) 1 and 100,   (iii) $\frac{1}{2}$ and 2,

(iv) $\frac{3}{4}$ and $\frac{4}{3}$,   (v) 1 and 2,   (vi) 2 and 4.

To find the mean proportional between two given straight lines.

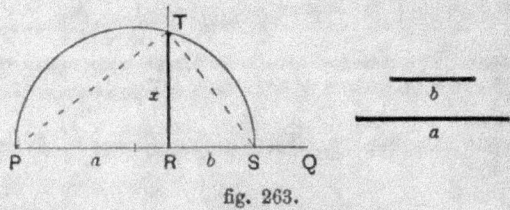

fig. 263.

Let a, b be the two given straight lines.

Construction   Draw a straight line PQ.

From PQ cut off PR $= a$, and RS $= b$.

On PS as diameter describe a semicircle.

Through R draw RT $\perp$ to PS to cut the semicircle at T.

Then RT (x) is the mean proportional between a, b.

Proof   Join PT, TS.

$\triangle^s$ PRT, TRS are equiangular.   (Why?)

$\therefore$ RP : RT $=$ RT : RS,

$\therefore$ $a : x = x : b$,

$\therefore$ x is the mean proportional between a and b.

†Ex. **1358.** Prove the above construction by completing the circle, and producing TR to meet the circle in T'.

Ex. **1359.** (On inch paper.) Find graphically the mean proportionals between (i) 1 and 4, (ii) 1 and 3, (iii) 1·5 and 2·5, (iv) 1·3 and 1·7.

Check by calculation.

Ex. **1360.** (On inch paper.) Find the square roots of (i) 2, (ii) 3, (iii) 6, (iv) 7. [Find the mean proportionals between (i) 1 and 2, (iii) 2 and 3.]

To describe a square equivalent to a given rectilinear figure.

Construction   (i)   Reduce the figure to a triangle (see p. 124).

(ii)   Convert the triangle into a rectangle.

(iii)   Find the mean proportional between the sides
of the rectangle.

This will be the side of the required square.

Proof If a, b are the sides of the rectangle, x the side of the equivalent square, then

$$\text{area of rectangle} = ab = x^2.$$

Ex. 1361. (On inch paper.) Find the side of the square equivalent to the triangles whose angular points are $(1, 0)$, $(5, 0)$, $(4, 3)$.

Ex. 1362. Construct a square equivalent to a regular hexagon of side 2 in.; measure the side of the square.

†**Ex. 1363.** In fig. 263, prove that (i) $PT^2 = PR \cdot PS$, (ii) $ST^2 = SR \cdot SP$.

†**Ex. 1364.** Prove Pythagoras' theorem by drawing the altitude to the hypotenuse and using similar triangles (see Ex. 1363).

Constructions by means of similarity. This method is illustrated by the following exercises.

¶ **Ex. 1365.** Show how to construct a triangle given its angles and the length of a median (say 4 inches).

[Construct a triangle with the given angles. Suppose that the corresponding median in this triangle is 3 inches. Magnify the whole figure in the ratio $3 : 4$; the result is the figure required.]

Ex. 1366. Show how to describe a triangle, having given its angles and its perimeter.

Ex. 1367. Show how to describe a triangle, having given its angles and the difference of two of its sides.

Ex. 1368. Show how to inscribe in a given triangle a triangle which has its sides parallel to the sides of another given triangle.

Ex. 1369. Show how to inscribe a square in a given triangle.

Ex. 1370. Show how to inscribe a square in a given sector of a circle.

Ex. 1371. Show how to inscribe an equilateral triangle in a given triangle.

Ex. 1372. Show how to describe a circle to touch two given straight lines and pass through a given point.

Ex. 1373. Show how to inscribe a regular octagon in a given square.

Theorem 8 (i).

The internal bisector of an angle of a triangle divides the opposite side internally in the ratio of the sides containing the angle.

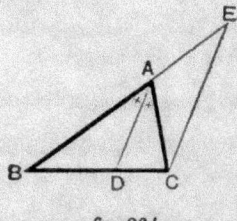

fig. 264.

Data	ABC is a triangle,
	AD bisects ∠ BAC internally and cuts BC at D.

To prove that $\dfrac{DB}{DC} = \dfrac{AB}{AC}$.

Construction   Through C draw CE ∥ to DA to cut BA produced at E.

Proof   Since DA is ∥ to CE,

$$\therefore \frac{DB}{DC} = \frac{AB}{AE}.$$ IV. 1.

[It remains to prove that AE = AC.]

  ∵ DA is ∥ to CE,

  ∴ ∠ BAD = corresp. ∠ AEC,   I. 5.

  and ∠ DAC = alt. ∠ ACE.   I. 5.

  But ∠ BAD = ∠ DAC,   *Data*

  ∴ ∠ AEC = ∠ ACE,

  ∴ AE = AC,   I. 13.

$$\therefore \frac{DB}{DC} = \frac{AB}{AC}.$$

Q. E. D.

†Ex. **1374.**   State and prove the converse of this theorem.
[Use *reductio ad absurdum.*]

THEOREM 8 (ii).

The external bisector of an angle of a triangle divides the opposite side externally in the ratio of the sides containing the angle.

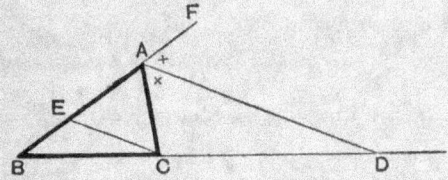

fig. 265.

Data ABC is a triangle,

AD bisects ∠ BAC externally (i.e. AD bisects ∠ CAF) and cuts BC produced at D.

To prove that
$$\frac{DB}{DC} = \frac{AB}{AC}.$$

Construction Through C draw CE ∥ to DA to cut BA at E.

Proof Since DA is ∥ to CE,

$$\therefore \frac{DB}{DC} = \frac{AB}{AE}. \qquad \text{IV. 1.}$$

[It remains to prove that AE = AC.]

$$\begin{cases} \because \text{ DA is} \parallel \text{to CE,} \\ \therefore \ \angle \text{FAD} = \text{corresp.} \ \angle \text{AEC,} \qquad \text{I. 5.} \\ \text{and } \angle \text{DAC} = \text{alt.} \ \angle \text{ACE.} \qquad \text{I. 5.} \\ \text{But } \angle \text{FAD} = \angle \text{DAC,} \qquad \text{Data} \\ \therefore \ \angle \text{AEC} = \angle \text{ACE,} \\ \therefore \ \text{AE} = \text{AC,} \qquad \text{I. 13.} \end{cases}$$

$$\therefore \frac{DB}{DC} = \frac{AB}{AC}.$$

Q. E. D.

NOTE. There is a very close analogy between theorems 8 (i) and 8 (ii) ; notice that the proofs are nearly identical.

†Ex. 1375. State and prove the converse of this theorem.

Ex. **1376.** In a △ABC, BC=3·5 in., CA=3 in., AB=4 in. and the internal bisector of ∠A cuts BC at D; calculate BD, DC.

Ex. **1377.** The internal bisector of ∠B of △ABC cuts the opposite side in E; find EA, EC when BC=8·9 cm., CA=11·5 cm., AB=4·7 cm.

Ex. **1378.** In a △ABC, BC=3·5 in., CA=3 in., AB=4 in. and the external bisector of ∠A cuts the base produced at D; calculate BD, DC.

Ex. **1379.** Repeat Ex. 1378 with

      (i) BC=5·2 in.,   CA=4·1 in., AB=4·5 in.,

      (ii) BC=11·5 cm., CA=4·7 cm., AB=8·9 cm.

†Ex. **1380.** The base BC of a triangle ABC is bisected at D. DE, DF bisect ∠ˢ ADC, ADB, meeting AC, AB in E, F. Prove that EF is ∥ to BC.

†Ex. **1381.** A straight line cuts a pair of intersecting straight lines at X, Y; it cuts the bisectors of the angles between the lines at H, K. Prove that XY is divided internally and externally in the same ratio.

†Ex. **1382. A point P moves so that the ratio of its distances from two fixed points Q, R is constant; prove that the locus of P is a circle. (The Circle of Apollonius.)**

[Draw the internal and external bisectors of ∠P.]

†Ex. **1383.** O is a point inside a triangle ABC. The bisectors of ∠ˢ BOC, COA, AOB meet BC, CA, AB in P, Q, R respectively. Prove that

$$\frac{BP}{PC} \times \frac{CQ}{QA} \times \frac{AR}{RB} = 1.$$

MISCELLANEOUS EXERCISES.

†Ex. **1384.** One of the parallel sides of a trapezium is double the other; show that the diagonals trisect one another.

†Ex. **1385.** Find the locus of a point which moves so that the ratio of its distances from two intersecting straight lines is constant.

Ex. **1386.** Show how to draw through a given point within a given angle a straight line to be terminated by the arms of the angle, and divided in a given ratio (say $\frac{2}{3}$) at the given point.

†Ex. **1387. Prove that two medians of a triangle trisect one another. Hence prove that the three medians pass through one point.**

†Ex. **1388.** The sides AC, BD of two triangles ABC, DBC on the same base BC and between the same parallels meet at E; prove that a parallel to BC through E, meeting AB, CD, is bisected at E.

Ex. **1389.** Show how to divide a parallelogram into five equivalent parts by lines drawn through an angular point.

†Ex. **1390.** Show how to divide a given line into two parts such that their mean proportional is equal to a given line. Is this always possible?

Ex. **1391.** Show how to construct a rectangle equivalent to a given square, and having its perimeter equal to a given line. [See Ex. 1390.]

†Ex. **1392.** A common tangent to two circles cuts the line of centres externally or internally in the ratio of the radii.

Ex. **1393.** Show how to construct on a given base a triangle having given the vertical angle and the ratio of the two sides.

Ex. **1394.** Show how to construct a triangle having given the vertical angle, the ratio of the sides containing the angle, and the altitude drawn to the base.

†Ex. **1395.** TP, TQ are tangents to a circle whose centre is C, CT cuts PQ in N; prove that $CN \cdot CT = CP^2$.

†Ex. **1396.** In fig. 259, prove that $PB \cdot PC : PA \cdot PD = BC^2 : AD^2$.

†Ex. **1397.** ABCDE is a regular pentagon; BE, AD intersect at F; prove that EF is a third proportional to AD, AE.

†Ex. **1398.** In two circles ABC, DEF, ∠BAC = ∠EDF, prove that the ratio of the chords BC, EF is equal to the ratio of the diameters of the circles.

†Ex. **1399.** In fig. 261, PQ is drawn parallel to AD to meet BC produced in Q; prove that PQ is a mean proportional between QB, QC.

†Ex. **1400.** The angle BAC of a △ABC is bisected by AD, which cuts BC in D; DE, DF are drawn parallel to AB, AC and cut AC, AB at E, F respectively. Prove that BF : CE = AB² : AC².

†Ex. **1401.** ABC is a triangle right-angled at A; AD is drawn perpendicular to BC and produced to E so that DE is a third proportional to AD, DB; prove that △ABD = △CDE, and △ABD is a mean proportional between △ˢ ADC, BDE.

†Ex. **1402.** Two circles touch externally at P; Q, R are the points of contact of one of their common tangents. Prove that QR is a mean proportional between their diameters. [Draw the common tangent at P, let it cut QR at S; join S to the centres of the two circles.]

†Ex. **1403.** Two church spires stand on a level plain; a man walks on the plain so that he always sees the tops of the spires at equal angles of elevation. Prove that his locus is a circle.

†Ex. **1404.** The rectangle contained by two sides of a triangle is equal to the square on the bisector of the angle between those sides together with the rectangle contained by the segments of the base. [See Ex. 1337.]

†Ex. **1405.** The tangent to a circle at P cuts two parallel tangents at Q, R; prove that the rectangle QP . PR is equal to the square on a radius of the circle.

†Ex. **1406.** ABCD is a quadrilateral. If the bisectors of ∠ˢ A, C meet on BD, then the bisectors of ∠ˢ B, D meet on AC.

Ex. **1407.** Prove the validity of the following method of **solving a quadratic equation** graphically :—

Suppose that $ax^2 + bx + c = 0$ is the equation; on squared paper, mark the origin, from OX cut off OP = a, from P draw a perpendicular PQ upwards of length b, from Q draw to the left QR = c (regard must be paid to the signs of a, b, c; e.g. if b is negative PQ will be drawn downwards); on OR describe a semicircle cutting PQ at S, T; the roots of the equation are $-\dfrac{PS}{OP}$ and $-\dfrac{PT}{OP}$. [Consider △ˢ OPS, SQR.]

Ex. **1408**. Solve the following equations graphically as in Ex. 1407, and check by calculation :— (i) $2x^2 + 5x + 1 = 0$, (ii) $x^2 + 3x - 2 = 0$, (iii) $2x^2 - x + 1 = 0$.

†Ex. **1409**. A straight line AB is divided internally at C; equilateral triangles ACD, CBE are described on the same side of AB; DE and AB produced meet at F. Prove that FB : BC=FC : CA.

†Ex. **1410**. ABC is an equilateral triangle and from any point D in AB straight lines DK and DL are drawn parallel to BC and AC respectively. Find the ratio of the perimeter of the parallelogram DLCK to the perimeter of the triangle ABC.

†Ex. **1411**. If from each of the angular points of a quadrilateral perpendiculars are let fall upon the diagonals, the feet of these perpendiculars are the angular points of a similar quadrilateral.

†Ex. **1412**. ABCD is a parallelogram, P is a point in AC produced; BC, BA are produced to cut the straight line through P and D in Q, R respectively. Prove that PD is a mean proportional between PQ and PR.

†Ex. **1413**. ABCD is a quadrilateral inscribed in a circle of which AC is a diameter; from any point P in AC, PQ and PR are drawn perpendicular to CD and AB respectively. Prove that DQ : PR=DC : BC.

†Ex. **1414**. Two circles ABC, ADE touch internally at A; through A straight lines ABD, ACE are drawn to cut the circles. Prove that AB . DE=AD . BC.

†Ex. **1415**. In the sides AD, CB of a quadrilateral ABCD points P, Q are taken so that AP : PD=CQ : QB. Prove that △ADQ+ △BPC=ABCD.

†Ex. **1416**. Find a point P in the arc AB of a circle such that chord AP is three times the chord PB.

†Ex. **1417**. Show how to draw through a given point D in the side AB of a triangle ABC a straight line DPQ cutting AC in P and BC produced in Q so that PQ is twice DP.

†Ex. **1418**. Show how to draw through a given point O a straight line to cut two given straight lines in P and Q respectively so that OP : PQ is equal to a given ratio.

†Ex. **1419**. O is a fixed point inside a circle, P is a variable point on the circle; what is the locus of the mid-point of OP?

†Ex. **1420**. ABCD is a quadrilateral; show how to draw through A, B parallel straight lines to cut CD in X, Y so that CX=DY. [X and Y are both to be between C and D, or one in CD produced and the other in DC produced.]

†Ex. **1421.** Show how to construct a triangle having given the lengths of two of its sides and the length of the bisector (terminated by the base) of the angle between them.

†Ex. **1422.** From any point X in a chord PR of a circle, XY is drawn perpendicular to the diameter PQ, prove that PX : PY = PQ : PR.

†Ex. **1423.** Through the vertex A, of a triangle ABC, DAE is drawn parallel to BC and AD is made equal to AE; CD cuts AB at X and BE cuts AC at Y; prove XY parallel to BC.

†Ex. **1424.** ABCD is a parallelogram; a straight line through A cuts BD in O, BC in P, DC in Q. Prove that AO is a mean proportional between OP and OQ.

†Ex. **1425.** A triangle PQR is inscribed in a circle and the tangent to the circle at the other end of the diameter through P cuts the sides PQ, PR produced at H, K respectively; prove that the △s PKH, PQR are similar.

†Ex. **1426.** Two circles ACB, ADB intersect at A, B; AC, AD touch the circles ADB, ACB respectively at A; prove that AB is a mean proportional between BC and BD.

†Ex. **1427.** A variable circle moves so as always to touch two fixed circles; prove that the straight line joining the points of contact cuts the line of centres of the fixed circle in one of two fixed points.

†Ex. **1428.** ABC is an equilateral triangle and D is any point in BC. On BC produced points E and F are taken such that AB bisects the angle EAD and AC bisects the angle DAF. Show that the triangles ABE and ACF are similar and that BE . CF = BC².

†Ex. **1429.** (i) In a △ABC, AB = ½AC; CX is drawn perpendicular to the internal bisector of the ∠BAC; prove that AX is bisected by BC.

(ii) State and prove an analogous theorem for the external bisector of the ∠BAC.

†Ex. **1430.** Two circles touch one another externally at A; BA and AC are diameters of the circles; BD is a chord of the first circle which touches the second at X, and CE is a chord of the second which touches the first at Y. Prove that BD . CE = 4DX . EY.

†Ex. **1431.** Two straight lines AOB, COD intersect at O; prove that, if OA : OB = OC : OD, then the △s AOD, BOC are equivalent.

†Ex. **1432.** The sides AB, AD of the rhombus ABCD are bisected in E, F respectively. Prove that the area of the triangle CEF is three-eighths of the area of the rhombus.

†Ex. **1433.** ABC is a triangle right-angled at A, the altitude AD is produced to E so that DE is a third proportional to AD, DC; prove that △s BDE, ADC are equal in area.

†Ex. **1434.** Two circles ABC, AB'C', whose centres are O and O', touch externally at A; BAB' is a straight line; prove that the triangles OAB', O'AB are equal in area.

†Ex. **1435.** ABC is a triangle, and BC is divided at D so that $BD^2 = BC . DC$. A line DE parallel to AC meets AB in E. Show that the triangles DBE, ACD are equal in area.

†Ex. **1436.** PA, PB are the two tangents from P to a circle whose centre is O; prove that $\triangle PAB : \triangle OAB = PA^2 : OA^2$.

†Ex. **1437.** Two triangles ABC, DEF have ∠A and ∠D supplementary and the sides about these angles proportional, prove that the ratio of the areas of these triangles is equal to $AB^2 : DE^2$.

†Ex. **1438.** Through the vertices of a triangle ABC, parallel straight lines are drawn to meet the opposite sides of the triangle in points α, β, γ; prove that $\triangle \alpha\beta\gamma = 2 \triangle ABC$.

Ex. **1439.** Through the vertices A, B, C of an equilateral triangle straight lines are drawn perpendicular to the sides AB, BC, CA respectively, so as to form another equilateral triangle. Compare the areas of the two triangles.

†Ex. **1440.** A square BCDE is described on the base BC of a triangle ABC, and on the side opposite to A. If AD, AE cut BC in F, G respectively, prove that FG is the base of a square inscribed in the triangle ABC.

†Ex. **1441.** Prove that the rectangle contained by the hypotenuses of two similar right-angled triangles is equal to the sum of the rectangles contained by the other pairs of corresponding sides.

†Ex. **1442.** The sides AB, AC of a triangle are bisected at D and E respectively; prove that, if the circle ADE intersect the line BC, and P be a point of intersection, then AP is a mean proportional between BP and CP.

†Ex. **1443.** Circles are described on the sides of a right-angled triangle ABC as diameters, and through the right angle A a straight line APQR is drawn cutting the three circles in P, Q, R respectively. Show that AP = QR.

†Ex. **1444.** The bisector of the angle BAC of a triangle ABC meets the side BC at D. The circle described about the triangle BAD meets CA again at E, and the circle described about the triangle CAD meets BA again at F. Show that BF is equal to CE.

†Ex. **1445.** D, E, F are points in the sides BC, CA, AB of a △ABC such that AD=BE=CF. From any point O within the △ABC, OP, OQ, OR are drawn parallel to AD, BE, CF to meet BC, CA, AB in P, Q, R respectively. Show that OP+OQ+OR=AD.

†Ex. **1446**. ABCD is a quadrilateral with the angles at A and C right angles. If BK and DN are drawn perpendicular to AC, prove that AN=CK.

†Ex. **1447.** The angle BAC of a triangle is bisected by a straight line which meets the base BC in D; a straight line drawn through D at right angles to AD meets AB in E and AC in F. Prove that EB : CF=BD : DC.

†Ex. **1448.** If the tangents at the ends of one diagonal of a cyclic quadrilateral intersect on the other diagonal produced, the rectangle contained by one pair of opposite sides is equal to that contained by the other pair.

†Ex. **1449.** Two circles ABC, APQ (of which APQ is the smaller) touch internally at A; BC a chord of the larger touches the smaller at R; AB, AC cut the circle APQ at P and Q respectively. Prove that AP : AQ=BR : RC.

†Ex. **1450.** AB is a fixed chord of a circle; CD is the diameter perpendicular to AB; P is a variable point on the circle; AP, BP cut CD (produced if necessary) in X, Y; if O is the centre of the circle, prove that OX . OY is constant.

†Ex. **1451.** Any point P is taken within a parallelogram ABDC; PM and PN are drawn respectively parallel to the sides AC and AB and terminated by AB and AC; NP produced meets BD in E; AE is joined meeting PM in P'; P'Q is drawn parallel to AB meeting the diagonal AD in Q. Prove that AQ : AD=parallelogram AMPN : parallelogram ABDC.

†Ex. **1452.** A straight line HK is drawn parallel to the base BC of a triangle ABC to cut AB, AC in H, K respectively; BK, HC intersect at X; AX cuts HK, BC at Y, Z respectively. Prove that YX : XZ=AY : AZ.

†Ex. **1453.** ABCDEFG is a regular heptagon; BG cuts AC, AD in X, Y respectively; prove that AX . AC=AY . AD.

†Ex. **1454.** P, Q, R, S are four consecutive corners of a regular polygon; PR, QS intersect at X; prove that QR is a mean proportional between PR and RX.

†Ex. **1455.** Two straight lines BGE, CGF intersect at G so that GE=½BE and GF=⅓CF; BF and CE are produced to meet at A; prove that BF=FA and CE=EA.

†Ex. **1456.** ABCDEF is a hexagon with its opposite sides parallel, CF is parallel to AB (and DE), and AD is parallel to BC (and EF); prove that BE must be parallel to CD (and AF).

APPENDIX.

THEOREM I. 1.

If a straight line stands on another straight line, the sum of the two angles so formed is equal to two right angles.

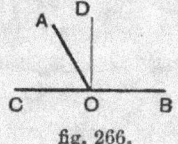

fig. 266.

Data   The st. line AO meets the st. line BC at O.

To prove that   ∠ BOA + ∠ AOC = 2 rt. ∠ s.

Construction   Draw OD to represent the line through O perpendicular to BC.

Proof   ∠ BOA = ∠ BOD + ∠ DOA,

∠ AOC = ∠ DOC − ∠ DOA,

∴ ∠ BOA + ∠ AOC = ∠ BOD + ∠ DOC

= 2 rt. ∠ s.   *Constr.*

Q. E. D.

Cor.   If any number of straight lines meet at a point, the sum of all the angles made by consecutive lines is equal to four right angles.

THEOREM I. 2.

[CONVERSE OF THEOREM I. 1.]

If the sum of two adjacent angles is equal to two right angles, the exterior arms of the angles are in the same straight line.

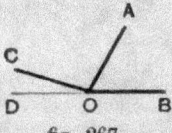

fig. 267.

Data   The sum of the adjacent ∠ s BOA, AOC = 2 rt. ∠ s.

To prove that   BOC is a straight line.

Construction   Produce BO to D.

Proof   Since AO meets the st. line BD at O,

∴ ∠ BOA + ∠ AOD = 2 rt. ∠ s.   I. 1.

But ∠ BOA + ∠ AOC = 2 rt. ∠ s,   *Data*

∴ ∠ BOA + ∠ AOD = ∠ BOA + ∠ AOC,

∴ ∠ AOD = ∠ AOC,

∴ OC coincides with OD.

Now BOD is a st. line,   *Constr.*

∴ BOC is a st. line.

THEOREM I. 3.

If two straight lines intersect, the vertically opposite angles are equal.

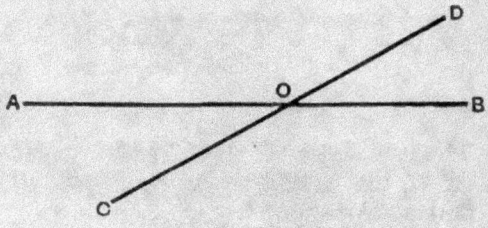

fig. 268.

Data The two st. lines AOB, COD intersect at O.

To prove that   ∠ AOD = vert. opp. ∠ BOC,

                         ∠ AOC = vert. opp. ∠ BOD.

Proof Since st. line OD stands on st. line AB,

                ∴ ∠ AOD + ∠ DOB = 2 rt. ∠ s,   I. 1.

and since st. line OB stands on st. line CD,

                ∴ ∠ DOB + ∠ BOC = 2 rt. ∠ s,   I. 1.

          ∴ ∠ AOD + ∠ DOB = ∠ DOB + ∠ BOC,

                ∴ ∠ AOD = ∠ BOC.

          Sim^ly ∠ AOC = ∠ BOD.   Q. E. D.

THEOREM I. 4.

(1) **When a straight line cuts two other straight lines, if a pair of alternate angles are equal, then the two straight lines are parallel.**

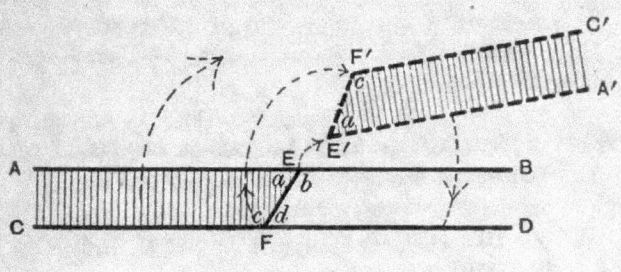

fig. 269.

(1) *Data* The st. line EF cuts the two st. lines AB, CD at E, F, forming the ∠ s a, b, c, d; and ∠ a = alternate ∠ d.

To prove that   AB, CD are parallel.

Proof

$$\left\{ \begin{array}{l} \angle a + \angle b = 2 \text{ rt. } \angle \text{s}, \\ \angle c + \angle d = 2 \text{ rt. } \angle \text{s}, \\ \therefore \angle a + \angle b = \angle c + \angle d. \\ \text{But } \angle a = \angle d. \\ \therefore \angle b = \angle c. \end{array} \right.$$

I. 1.
I. 1.

Data

17—2

Take up the part AEFC, call it A′E′F′C′; and, turning it round in its own plane, apply it to the part DFEB so that E′ falls on F and E′A′ along FD.

$$\because \angle a = \angle d,$$ *Data*

∴ E′F′ falls along FE,

and ∵ E′F′ = FE (being the same line),

∴ F′ falls on E ;

again ∵ ∠ c = ∠ b, *Proved*

∴ F′C′ falls along EB.

Now if EB and FD meet when produced towards B and D, F′C′ and E′A′ must also meet when produced towards C′ and A′, i.e. FC and EA must also meet when produced towards C and A.

∴ if AB, CD meet when produced in one direction, they will also meet when produced in the other direction ; but this is impossible, for two st. lines cannot enclose a space.

∴ AB, CD cannot meet however far they are produced in either direction.

∴ AB and CD are parallel. Q. E. D.

When a straight line cuts two other straight lines, if

　(2)　a pair of corresponding angles are equal,

or (3)　a pair of interior angles on the same side of the cutting line are together equal to two right angles,

then the two straight lines are parallel.

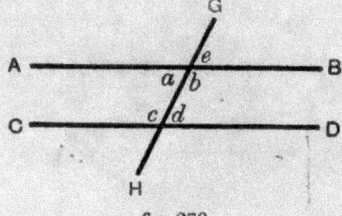

fig. 270.

(2) *Data*   The st. line GH cuts the two st. lines AB, CD forming the $\angle$s a, b, c, d, e.

$$\angle e = \text{corresp. } \angle d.$$

To prove that   AB, CD are parallel.

Proof $\angle e = \text{vert. opp. } \angle a$.   I. 3.

But $\angle e = \angle d$,   *Data*

$\therefore \angle a = \angle d$,

and these are alternate angles,

$\therefore$ AB, CD are parallel.   by (1).

(3) *Data* $\angle b + \angle d = 2 \text{ rt. } \angle \text{s}$.

To prove that   AB, CD are parallel.

Proof $\angle b + \angle a = 2 \text{ rt. } \angle \text{s}$.   I. 1.

But $\angle b + \angle d = 2 \text{ rt. } \angle \text{s}$.   *Data*

$\therefore \angle b + \angle a = \angle b + \angle d$,

$\therefore \angle a = \angle d$,

and these are alternate angles,

$\therefore$ AB, CD are parallel.   by (1).

Q. E. D.

Cor.   If each of two straight lines is perpendicular to a third straight line, the two straight lines are parallel to one another.

Playfair's Axiom.   Through a given point one straight line, and one only, can be drawn parallel to a given straight line.

THEOREM I. 5.
[CONVERSE OF THEOREM I. 4.]

If a straight line cuts two parallel straight lines,

(1) alternate angles are equal,

(2) corresponding angles are equal,

(3) the interior angles on the same side of the cutting line are together equal to two right angles.

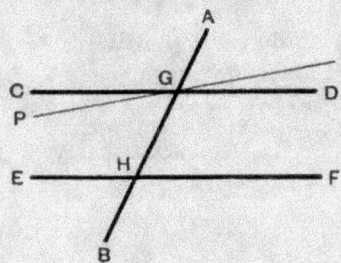

fig. 271.

Data   AB cuts the parallel st. lines CD, EF at G, H.

To prove that   (1)   ∠ CGH = alt. ∠ GHF,

(2)   ∠ AGD = corresp. ∠ GHF,

(3)   ∠ DGH + ∠ GHF = 2 rt. ∠ s.

(1) *Construction*   If ∠ CGH is not equal to ∠ GHF,
suppose GP drawn so that ∠ PGH = ∠ GHF.

Proof   ∵ ∠ PGH = alt. ∠ GHF,

∴ PG is ‖ to EF.   I. 4.

∴ the two straight lines PG, CG which pass through the point G are both ‖ to EF.

But this is impossible.   *Playfair's Axiom*

∴ ∠ CGH cannot be unequal to ∠ GHF,

∴ ∠ CGH = ∠ GHF.

(2)   Since, by (1), ∠ CGH = ∠ GHF

and ∠ CGH = vert. opp. ∠ AGD,

∴ ∠ AGD = ∠ GHF.

(3)   Since GH stands on CD,

                    ∴ ∠ DGH + ∠ CGH = 2 rt. ∠ s,   I. 1.

                    and, by (1), ∠ CGH = ∠ GHF,

                    ∴ ∠ DGH + ∠ GHF = 2 rt. ∠ s.   Q. E. D.

THEOREM I. 6.

Straight lines which are parallel to the same straight line are parallel to one another.

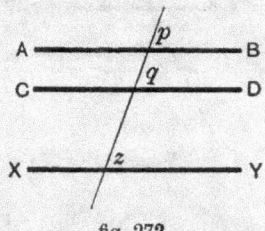

fig. 272.

Data   AB, CD are each ∥ to XY.

To prove that   AB is ∥ to CD.

Construction   Draw a st. line cutting AB, CD, XY and forming
        with them corresponding ∠ s p, q, z respectively.

Proof   ∵ AB is ∥ to XY,

                    ∴ ∠ p = corresp. ∠ z.   I. 5.

                    Again ∵ CD is ∥ to XY,

                    ∴ ∠ q = corresp. ∠ z,   I. 5.

                         ∴ ∠ p = ∠ q.

        Now these are corresponding angles,

                    ∴ AB is ∥ to CD.   I. 4.

                                        Q. E. D.

Theorem I. 7.

If straight lines are drawn from a point parallel to the arms of an angle, the angle between those straight lines is equal or supplementary to the given angle.

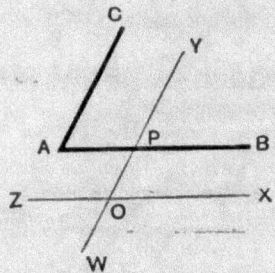

fig. 273.

Data   BAC is an angle.

From O, OX is drawn ‖ to AB and in the same sense as AB, and OY is drawn ‖ to AC and in the same sense as AC; XO, YO are produced to Z, W respectively.

To prove that   ∠ XOY = ∠ ZOW = ∠ BAC,

            ∠ YOZ = ∠ WOX = supplement of ∠ BAC.

Proof   Let WY cut AB at P,

            then ∠ XOY = corresp. ∠ BPY,   I. 5.

            and ∠ BAC = corresp. ∠ BPY,   I. 5.

                ∴ ∠ XOY = ∠ BAC.

            But ∠ ZOW = vert. opp. ∠ XOY,

                ∴ ∠ ZOW = ∠ BAC.

        Again ∠ YOZ = ∠ XOW = supplement of ∠ XOY

                        = supplement of ∠ BAC.

                                        Q. E. D.

THEOREM I. 8.

The sum of the angles of a triangle is equal to two right angles.

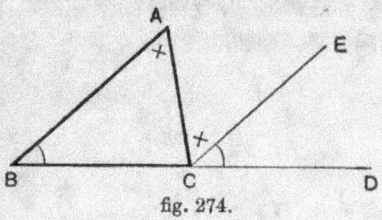

fig. 274.

Data	ABC is a triangle.
To prove that	$\angle A + \angle B + \angle ACB = 2$ rt. $\angle$ s.
Construction	Produce BC to D.
	Through C draw CE ∥ to BA.
Proof	Since AC cuts the ∥s AB, CE,

$\therefore \angle A =$ alt. $\angle$ ACE.

And since BC cuts the ∥s AB, CE,

$\therefore \angle B =$ corresp. $\angle$ ECD,

$\therefore \angle A + \angle B = \angle ACE + \angle ECD.$

Add $\angle$ ACB to each side,

$\therefore \angle A + \angle B + \angle ACB = \angle ACB + \angle ACE + \angle ECD$

$= 2$ rt. $\angle$ s   (for BCD is a st. line),

$\therefore$ sum of $\angle$ s of $\triangle$ ABC $= 2$ rt. $\angle$ s.

Q. E. D.

Cor. 1. If one side of a triangle is produced, the exterior angle so formed is equal to the sum of the two interior opposite angles. (Proof as above.)

Cor. 2. If two triangles have two angles of the one equal to two angles of the other, each to each, then the third angles are also equal.

Cor. 3. The sum of the angles of a quadrilateral is equal to four right angles. (Draw a diagonal.)

THEOREM I. 9.

If the sides of a convex polygon are produced in order, the sum of the angles so formed is equal to four right angles.

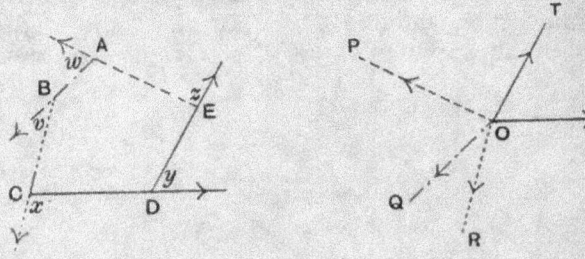

fig. 275.

Data   ABCDE is a convex polygon; its sides are produced in order and form the exterior angles w, v, x, y, z.

To prove that $\angle w + \angle v + \angle x + \angle y + \angle z = 4$ rt. $\angle$ s.

Construction   Through any point O draw OP, OQ, OR, OS, OT || to and in the same sense as EA, AB, BC, CD, DE respectively.

Proof   Since OP, OQ are respectively || to and in the same sense as EA, AB,

$$\therefore \angle w = \angle \text{POQ}, \qquad\qquad \text{I. 7.}$$
$$\text{Sim}^{\text{ly}} \angle v = \angle \text{QOR},$$
$$\angle x = \angle \text{ROS},$$
$$\angle y = \angle \text{SOT},$$
$$\angle z = \angle \text{TOP},$$
$$\therefore \angle w + \angle v + \angle x + \angle y + \angle z = \text{sum of } \angle \text{s at O}$$
$$= 4 \text{ rt. } \angle \text{ s.} \qquad \text{I. 1 } Cor.$$

Q. E. D.

COR.   The sum of the interior angles of any convex polygon together with four right angles is equal to twice as many right angles as the polygon has sides.

THEOREM I. 10.

If two triangles have two sides of the one equal to two sides of the other, each to each, and also the angles contained by those sides equal, the triangles are congruent.

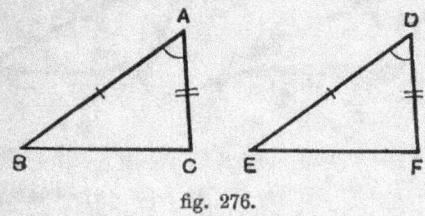

fig. 276.

Data   ABC, DEF are two triangles which have AB = DE, AC = DF, and included ∠ BAC = included ∠ EDF.

To prove that   △ ABC ≡ △ DEF.

Proof   Apply △ ABC to △ DEF so that A falls on D, and AB falls along DE.

∵ AB = DE,

∴ B falls on E.

Again ∵ ∠ BAC = ∠ EDF,

∴ AC falls along DF.

And ∵ AC = DF,

∴ C falls on F,

∴ △ ABC coincides with △ DEF,

∴ △ ABC ≡ △ DEF.

Q. E. D.

THEOREM I. 11.

If two triangles have two angles of the one equal to two angles of the other, each to each, and also one side of the one equal to the corresponding side of the other, the triangles are congruent.

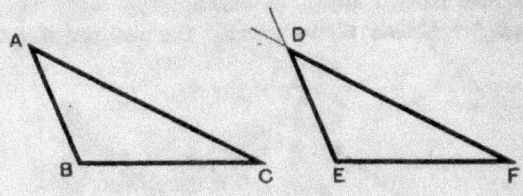

fig. 277.

Data   ABC, DEF are two triangles which have BC = EF and two angles of the one equal to the two corresponding angles of the other.

To prove that   △ABC ≡ △DEF.

Proof   Since two angles of △ABC are respectively equal to two angles of △DEF,

∴ the third angle of △ABC = the third angle of △DEF,

I. 8, *Cor.* 2.

∴ ∠A = ∠D,   ∠B = ∠E, and ∠C = ∠F.

Apply △ABC to △DEF so that B falls on E, and BC falls along EF.

∵ BC = EF,

∴ C falls on F.

Now ∠B = ∠E,

∴ BA falls along ED,

∴ A falls somewhere along ED or ED produced.

Again ∠C = ∠F,

∴ CA falls along FD,

∴ A falls somewhere along FD or FD produced,

∴ A falls on D,

∴ △ABC coincides with △DEF,

∴ △ABC ≡ △DEF.   Q. E. D.

Theorem I. 12.

If two sides of a triangle are equal, the angles opposite to these sides are equal. (Fig. 278.)

Data   ABC is a triangle which has AB = AC.

To prove that   ∠ C = ∠ B.

Construction   Draw AD to represent the bisector of ∠ BAC.
                    Let it cut BC at D.

Proof   In the △s ABD, ACD

$\left\{\begin{array}{l} AB = AC, \\ AD \text{ is common,} \\ \angle BAD = \angle CAD \text{ (included } \angle s), \end{array}\right.$ *Data*

                                     Constr.

∴ △ABD ≡ △ACD,   I. 10.

∴ ∠ B = ∠ C.   Q. E. D.

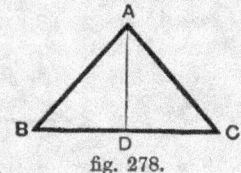

fig. 278.

Theorem I. 13.

[Converse of Theorem I. 12.]

If two angles of a triangle are equal, the sides opposite to these angles are equal. (Fig. 278.)

Data   ABC is a triangle which has ∠ B = ∠ C.

To prove that   AC = AB.

Construction   Draw AD to represent the bisector of ∠ BAC.
                    Let it cut BC at D.

Proof   In the △s ABD, ACD,

$\left\{\begin{array}{l} \angle B = \angle C, \\ \angle BAD = \angle CAD, \\ AD \text{ is common,} \end{array}\right.$ *Data*

                                     Constr.

∴ △ ABD ≡ △ ACD,   I. 11.

∴ AB = AC.   Q. E. D.

THEOREM I. 14.

If two triangles have the three sides of the one equal to the three sides of the other, each to each, the triangles are congruent.

Data   ABC, DEF are two triangles which have BC = EF, CA = FD, and AB = DE.

To prove that   △ABC ≡ △DEF.

Proof   Apply △ABC to △DEF so that B falls on E and BC falls along EF, but so that A and D are on opposite sides of EF; let A′ be the point on which A falls.   Join DA′.

Since BC = EF, C will fall on F.

CASE I.   *When* DA′ *cuts* EF.

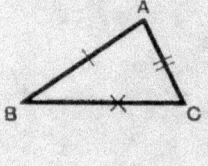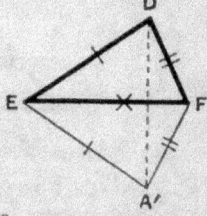

fig. 279.

In △EDA′,   ED = EA′ (i.e. BA),
  ∴ ∠ EA′D = ∠ EDA′.   I. 12.
In △FDA′,   FD = FA′ (i.e. CA),
  ∴ ∠ FA′D = ∠ FDA′,   I. 12.
∴ ∠ EA′D + ∠ FA′D = ∠ EDA′ + ∠ FDA′.
  i.e. ∠ EA′F = ∠ EDF,
  i.e. ∠ BAC = ∠ EDF,
∴ in △s ABC, DEF,
  ⎧   AB = DE,   *Data*
  ⎨   AC = DF,   *Data*
  ⎩ ∠ BAC = ∠ EDF (included ∠ s).   *Proved*
    ∴ △ABC ≡ △DEF.   I. 10.

Case ii.   *When DA' passes through one end of EF, say F.*

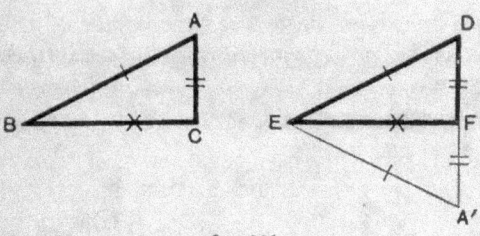

fig. 280.

In △ EDA', ED = EA' (i.e. BA),

∴ ∠ EA'D = ∠ EDA',   I. 12.

i.e. ∠ BAC = ∠ EDF,

∴ as in Case i.   △ ABC ≡ △ DEF.

Case iii.   *When DA' does not cut EF.*

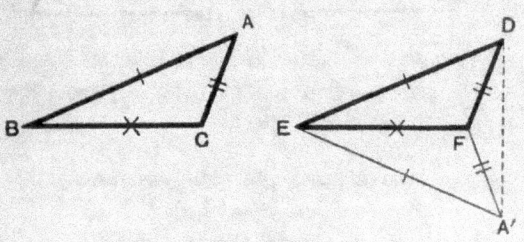

fig. 281.

As in Case i.   ∠ EA'D = ∠ EDA',

and ∠ FA'D = ∠ FDA',

∴ ∠ EA'D − ∠ FA'D = ∠ EDA' − ∠ FDA',

i.e. ∠ EA'F = ∠ EDF,

i.e. ∠ BAC = ∠ EDF,

∴ as in Case i.   △ ABC ≡ △ DEF.

Q. E. D.

THEOREM I. 15.

If two right-angled triangles have their hypotenuses equal, and one side of the one equal to one side of the other, the triangles are congruent.

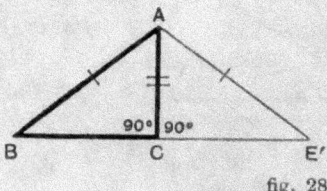 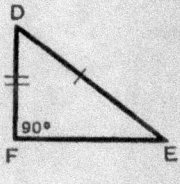

fig. 282.

Data   ABC, DEF are two triangles which have ∠s C, F right ∠s.
AB = DE, and AC = DF.

To prove that   △ ABC ≡ △ DEF.

Proof   Apply △ DEF to △ ABC so that D falls on A and DF along AC, but so that E and B are on opposite sides of AC ; let E′ be the point on which E falls.
    Since DF = AC, F will fall on C.
    Since ∠s ACB, ACE′ (i.e. DFE) are two rt. ∠s,   *Data*
        BCE′ is a st. line.   I. 2.
    ∴ ABE′ is a △.
    In this △, AB = AE′ (i.e. DE),   *Data*
        ∴ ∠ E′ = ∠ B.   I. 12.
    Now in the △s ABC, AE′C,
        { ∠ B = ∠ E′,   *Proved*
        { ∠ ACB = ∠ ACE′,   *Data*
        { AB = AE′,   *Data*
    ∴ △ ABC ≡ △ AE′C,   I. 11.
    ∴ △ ABC ≡ △ DEF.

                                    Q. E. D.

THEOREM I. 16.

If two sides of a triangle are unequal, the greater side has the greater angle opposite to it.

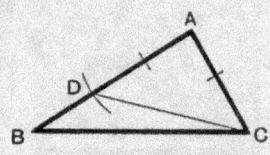

fig. 283.

Data   ABC is a triangle in which AB > AC.

To prove that   ∠ ACB > ∠ B.

Construction   From AB, the greater side, cut off AD = AC.
                      Join CD.

Proof   In △ ACD, AD = AC,
                    ∴ ∠ ACD = ∠ ADC.   I. 12.

But since the side BD of the △ DBC is produced to A,
              ∴ ext. ∠ ADC > int. opp. ∠ B,   I. 8, *Cor.* 1.
                  ∴ ∠ ACD > ∠ B.
             But ∠ ACB > its part ∠ ACD,
                  ∴ ∠ ACB > ∠ B.

                                             Q. E. D.

†Ex. **1457.** In a quadrilateral ABCD, AB is the shortest side and CD is the longest side; prove that ∠B > ∠D, and ∠A > ∠C.

†Ex. **1458.** Prove Theorem 16 by means of the following construction:—from AB cut off AD = AC, bisect ∠BAC by AE, join DE. (See fig. 284.)

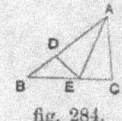

fig. 284.

G. S. S. G.   18

THEOREM I. 17.

[CONVERSE OF THEOREM I. 16.]

If two angles of a triangle are unequal, the greater angle has the greater side opposite to it.

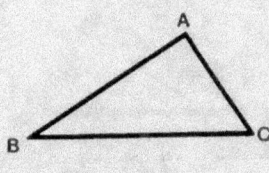

fig. 285.

Data   ABC is a triangle in which ∠ C > ∠ B.

To prove that   AB > AC.

Proof   Either   (i)   AB > AC,

          or   (ii)   AB = AC,

          or   (iii)   AB < AC.

If, as in (iii),   AB < AC,

          then   ∠ C < ∠ B,   I. 16.

          which is impossible.   *Data*

If, as in (ii),   AB = AC,

          then   ∠ C = ∠ B,   I. 12.

          which is impossible.   *Data*

          ∴ AB must be > AC.

Q. E. D.

†Ex. **1459.**   AD is drawn perpendicular to BC the opposite side of a triangle ABC; prove that AB > BD and AC > CD.

Hence show that AB + AC > BC.

[There will be two cases.]

†Ex. **1460.**   If the perpendiculars from B, C to the opposite sides of the triangle ABC intersect at a point X inside the triangle, and if AB > AC, prove that XB > XC.

THEOREM I. 18.

Any two sides of a triangle are together greater than the third side.

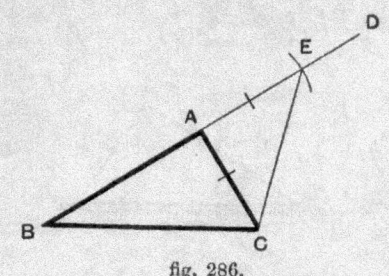

fig. 286.

Data   ABC is a triangle.

To prove that   (1)   BA + AC > BC,

(2)   CB + BA > CA,

(3)   AC + CB > AB.

(1)   *Construction*   Produce BA to D.

From AD cut off AE = AC.

Join CE.

Proof   In the △ AEC, AE = AC,   *Constr.*

∴ ∠ ACE = ∠ AEC,   I. 12.

But ∠ BCE > its part ∠ ACE,

∴ ∠ BCE > ∠ AEC,

∴ in the △ EBC, ∠ BCE > ∠ BEC,

∴ BE > BC,   I. 17.

i.e. BA + AE > BC,

∴ BA + AC > BC,   *Constr.*

(2)   Sim^ly CB + BA > CA,

(3)   and   AC + CB > AB.

Q. E. D.

†Ex. **1461.** If O is any point inside a triangle ABC, prove that BA + AC > BO + OC. [Produce BO to cut AC.]

18—2

THEOREM I. 21.

Of all the straight lines that can be drawn to a given straight line from a given point outside it, the perpendicular is the shortest.

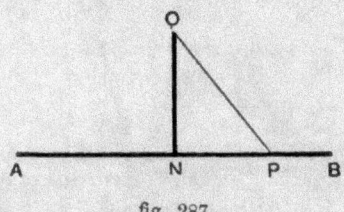

fig. 287.

Data   AB is a straight line and O a point outside it; ON is drawn ⊥ to AB meeting it in N.

To prove that   ON < any other st. line that can be drawn from O to AB.

Construction   Draw any other st. line from O to meet AB at P.

Proof   In the △ONP,
$$\angle N + \angle P < 2 \text{ rt. } \angle s, \qquad\qquad \text{I. 8.}$$
$$\text{and } \angle N = 1 \text{ rt. } \angle,$$
$$\therefore \ \angle P < 1 \text{ rt. } \angle,$$
$$\therefore \ \angle P < \angle N,$$
$$\therefore \ ON < OP, \qquad\qquad \text{I. 17.}$$

Sim^{ly} ON may be proved less than any other st. line drawn from O to meet AB.

∴ ON is the shortest of all such lines.

Q. E. D.

†Ex. **1462.** ABC, APQC, are a triangle and a convex quadrilateral on the same base AC, P and Q being inside the triangle; prove that the perimeter of the triangle is greater than that of the quadrilateral.
[Produce AP, PQ to meet BC and use I. 18.]

†Ex. **1463**. O is a point inside a triangle ABC; prove that ∠BOC > ∠BAC.

[Produce BO to cut AC.]

†Ex. **1464**. Two sides of a triangle are together greater than twice the median drawn through their point of intersection.

[Use the construction and figure of Ex. 285.]

†Ex. **1465**. O is a point inside a quadrilateral ABCD; prove that

$$OA + OB + OC + OD$$

cannot be less than AC + BD.

†Ex. **1466**. The sum of the distances of any point O from the vertices of a triangle ABC is greater than half the perimeter of the triangle.

[The perimeter of a figure is the sum of its sides. Apply I. 18 to △ˢ OBC, OCA, OAB in turn and add up the results.]

†Ex. **1467**. The sum of the distances from the vertices of a triangle of any point within the triangle is less than the perimeter of the triangle.

[Apply Ex. 1461 three times.]

Would this be true for a point outside the triangle?

†Ex. **1468**. The sum of the diagonals of a quadrilateral is greater than half its perimeter.

†Ex. **1469**. The sum of the diagonals of a quadrilateral is less than its perimeter.

†Ex. **1470**. A and B are two points on the same side of a straight line CD; find the point P in CD for which AP + PB is least. Give a proof. (See Ex. 635.)

EXAMINATION PAPERS.

CAMBRIDGE PREVIOUS EXAMINATION.

OCTOBER, 1911.

GEOMETRY. (2½ *hours*.)

[In all questions of Practical Geometry the methods of construction must be clearly indicated.]

1. Prove that, if two straight lines intersect, the vertically opposite angles are equal.

2. Show how to draw a straight line perpendicular to a given straight line at a given point in it, and justify your construction.

3. Prove that, if a straight line be drawn cutting two other straight lines so that a pair of alternate angles are equal, then the two straight lines are parallel.

4. Construct a quadrilateral ABCD of which AB=5 cm., BC=4 cm., CD=9 cm., DA=6 cm., and the sides AB and CD are parallel. Measure the lengths of its diagonals.

5. Prove that, in an obtuse-angled triangle, the square on the side subtending the obtuse angle is equal to the sum of the squares on the sides containing the obtuse angle together with twice the rectangle contained by one of those sides and the projection of the other side upon it.

6. Prove that one circle, and only one, can pass through any three points not in the same straight line.

Construct a triangle whose sides are 5 cm., 7 cm. and 10 cm.; describe a circle passing through its angular points, and measure the length of its radius.

7. Prove that the opposite angles of any quadrilateral inscribed in a circle are together equal to two right angles.

ABCD is a parallelogram and the circle described about the triangle ABC cuts CD, or CD produced, in E. Prove that AE is equal to AD.

8. Show how to draw a tangent to a circle from a given external point.

Draw a straight line AB whose length is 3·5 inches; with centre B and radius 2 inches draw a circle. Draw a tangent to the circle from the point A and measure its length.

9. In a triangle ABC the straight line which bisects the angle BAC cuts the side BC in the point D. Prove that the ratio of BD to DC is equal to the ratio of BA to AC.

AC is greater than AB and on AC a point E is taken so that AE is equal to AB, and through E a straight line EF is drawn parallel to AB to cut BC in F. Prove that BD is a mean proportional between DC and DF.

UNIVERSITY OF CAMBRIDGE.

PRELIMINARY LOCAL EXAMINATION. JULY, 1911.

GEOMETRY. (2 hours.)

Candidates can pass in Geometry by doing sufficiently well in Part I. Figures should be drawn accurately with a fairly hard pencil; but the letters attached to the figures, and the written answers, must be in **ink**. *In the constructions asked for in questions* 1, 2, 3, 6, 7, 8, *no explanations need be given;* **but all lines required in the constructions must be shown clearly.**

PART I.

1. Draw two lines AB, AC making with each other an angle of 55°. Make AB=3·5 inches and AC=5 inches. Without using a protractor or a set square bisect BC at X; draw the perpendicular XN from X to AB and measure its length in inches.

2. Construct a triangle PQM having PQ=9 cm., QM=7·8 cm., and MP=5·7 cm. Draw, without using a protractor, a line QS to bisect the angle PQM and to meet PM at S. Through S draw SA, SB parallel to MQ and PQ respectively to meet PQ and QM in A and B. Join AB and measure its length in centimetres.

3. The height of a cone is 10 cm., and the diameter of its base is 7 cm.; find by measurement the length of its slant side in centimetres.

4. AB and PQ are two lines cutting at X. Prove that the angles AXP, BXQ are equal.

5. Prove that the opposite sides and angles of a parallelogram ABXT are equal.

PQST is a parallelogram; the angles PQS, PTS are bisected by lines QK, TL meeting the diagonal PS in K and L respectively. Prove that QK=TL.

6. Prove that, if a triangle and a rectangle are on the same base and between the same parallels, then the area of the triangle is half that of the rectangle.

Find the area in square cms. of a triangle KLM in which KL=7 cm., LM=10 cm., and MK=6 cm.

PART II.

7. The line CD measures 12 cm.; with centre C describe a circle with radius 5 cm. Construct one of the tangents from D to the circle, and measure in centimetres the perpendiculars upon this from the points where the line CD cuts the circle.

8. Draw a triangle ABC such that BC=5 inches, the angle ABC=40°, and the angle ACB=65°. Construct the inscribed circle of the triangle, and measure its radius in inches.

9. AB, PQ are equal chords in a circle, whose centre is X. Prove that AB, PQ are equidistant from X.

A number of chords of length 8 cm. are placed in a circle of radius 6 cm. Show that their middle points all lie on a circle, and calculate the radius of this circle.

10. HKST is a quadrilateral inscribed in a circle; prove that the sum of the angles HKS, HTS is equal to two right angles.

ABC is an isosceles triangle having AB equal to AC; a line XY is drawn parallel to BC meeting AB, AC in X and Y respectively. Prove that X, Y, B, C lie on a circle.

UNIVERSITY OF CAMBRIDGE.

JUNIOR LOCAL EXAMINATION. JULY, 1911.

GEOMETRY. (2 *hours*.)

Candidates can pass in Geometry by doing sufficiently well in the A *part of this paper. Figures should be drawn accurately with a fairly hard pencil; but the letters attached to the figures, and the written answers, must be in* **ink.** *In questions* 1 *and* 6, **all the construction lines must be shown clearly.**

Answers to the questions marked A *are to be fastened together in one bundle; those to the questions marked* B *in another bundle.*

A.

A 1. In the triangles ABC, DEF, side BC=side EF, ∠ ABC=∠ DEF, ∠ ACB=∠ DFE. Prove that the triangles are congruent.

LMN is a triangle, and LO is drawn perpendicular to MN meeting it at O. It is found that LO=4 cm., LM=LN=5 cm. From these data construct the triangle LMN.

Having constructed the triangle LMN, take a point P in LM and a point Q in LN, making $\angle$ PNM $=\angle$ QMN. Then prove by geometrical reasoning that MQ=NP.

A 2. In the triangle ABC the angle A is a right angle. Prove that

$$BC^2 = AB^2 + AC^2.$$

The lengths of the three edges, terminating at one corner, of a rectangular block are 3, 4, and 5 inches respectively. Find correctly to the tenth of an inch the lengths of the diagonals of the different faces of the block.

A 3. Prove that if a straight line, drawn through the centre of a circle, bisects a chord of the circle, which is not a diameter, it cuts it at right angles.

Prove that if the line joining the middle points of two chords of a circle, which are not diameters, passes through the centre of the circle, the chords are parallel.

A 4. In the quadrilateral ABCD the opposite angles A and C are supplementary. Prove that the circle described to pass through the points A, B, and C also passes through D.

A parallelogram is inscribed in a circle. Prove that the angles of the parallelogram are right angles.

B.

B 5. Illustrate and explain by a figure the identity

$$a^2 - b^2 = (a+b)(a-b).$$

In the triangle ABC, the side AB is greater than the side AC. BC is bisected at O, and AD is drawn perpendicular to BC, cutting it at D. Prove that AB^2 - AC^2 = 4OB . OD.

B 6. From a point O are drawn two lines, one touching a circle at T, and the other cutting the same circle at M and N. Prove that

$$OT^2 = OM . ON.$$

Draw a straight line OBC, making OB=2·5 cm., OC=6·4 cm. Through O draw a line OA making the angle AOB=42°. Then draw a circle passing through B and C and touching OA. (Describe the steps of your construction.)

B 7. Prove that the external bisector of an angle of a triangle divides externally the side opposite the angle in the ratio of the sides containing the angle.

ABC is a triangle. O is the middle point of BC, and AO is produced to T. The lines bisecting internally the angles BOT, COT cut externally the sides AB, AC in D, E. Prove that DE is parallel to BC.

B 8. A point O within a triangle ABC is joined to each of the vertices, and OA, OB, OC are divided in X, Y, Z respectively, so that the ratios OX : OA, OY : OB, OZ : OC are equal. Prove that the triangles XYZ and ABC are similar.

If in the above AB = 5 in., BC = 12 in., AC = 13 in., and each of the equal ratios is one-third, compute the area of the triangle XYZ.

UNIVERSITY OF CAMBRIDGE.

SENIOR LOCAL EXAMINATION. JULY, 1911.

GEOMETRY. (2 hours.)

The answers to questions marked A and B are to be arranged and sent up to the Examiner in two separate bundles.

[N.B. *All construction lines should be clearly shown. Attention is called to the alternative questions* B ix, B x, B xi *at the end of the paper.*]

A.

A 1. Prove that if in the triangles ABC, DEF, AB = DE, AC = DF, and ∠A = ∠D, then the triangles are congruent.

X, Y, and Z are the middle points of the sides of an equilateral triangle PQR. Prove that the triangle XYZ is equilateral.

A 2. Find the locus of a point which is equidistant from two given intersecting straight lines.

In a given triangle ABC find points in BC and BC produced which are equidistant from AB and AC.

In what special case can only one such point be found?

A 3. Construct a quadrilateral ABCD in which AB = 7·2 cm., AD = 4 cm., CD = 9·6 cm., ∠A = 126°, and the diagonals AC and BD are perpendicular to each other. Find the area of the quadrilateral.

A 4. Prove that if the sides of a convex polygon are produced in order, the sum of the angles so formed is equal to four right angles.

Is it possible to make a pavement consisting of equal equilateral triangles, leaving no gaps?

Is it possible to do so with equal regular polygons of (i) four, (ii) five, or (iii) six sides? Give reasons for your answers.

B.

B 5. If two circles touch externally, prove that the straight line joining their centres passes through their point of contact.

Two circles touch externally at a point A. A common tangent to the two circles touches them at B and C. Prove that BAC is a right angle.

B 6. Draw a circle of radius 1·5 inches, touching each of two equal circles, of radius ·8 inch, whose centres are 2 inches apart.

Show that six circles can be drawn satisfying the given conditions.

B 7. From F, a point in the side PQ of the triangle PQR, a straight line is drawn parallel to QR, meeting PR in G. Prove that

$$PF : FQ = PG : GR.$$

A given point D lies between two given straight lines AB and AC. Find a construction for a line through D terminated by AB and AC, such that D is one of its points of trisection. Prove also that there are two such lines.

B 8. Explain, with proof, how to draw a perpendicular to a given plane from a given point outside it.

A sphere of radius 2 inches rests on three horizontal wires, forming a plane equilateral triangle of side 6 inches. Find the height of the top of the sphere above the plane of the wires.

[N.B. *One or more of the following questions* B ix, B x, B xi *may be taken instead of an equal number of the questions* B 6, B 7, B 8, *but lower marks will be assigned to them.*]

B ix. Construct a parallelogram whose diagonals include an angle of 76° and are 8·4 cm. and 5·6 cm. long. Measure its acute angle and find its area.

B x. ABC is a triangle having an acute angle at B. AD is drawn perpendicular to BC, produced if necessary. Prove that a square described on AC will be equal to the sum of the squares on AB, BC, less twice the rectangle contained by BC and BD.

State, without proof, the corresponding theorem when ∠B is obtuse.

B xi. Prove that opposite angles of any quadrilateral inscribed in a circle are supplementary.

P, Q, R are any points in the sides BC, CA, AB of a triangle. Prove that the circles circumscribing the triangles AQR, BRP, CPQ meet in a point.

OXFORD RESPONSIONS. SEPTEMBER, 1911.

GEOMETRY. (2½ hours.)

1. Draw an equilateral triangle each of whose sides is 5 cm. Bisect BC at D and AC at E. Join DE and measure the length of DE.

2. Draw a triangle with sides 6 cm., 7 cm., and base 8 cm. long. On a base 2 inches long draw another triangle equiangular to the first and measure its sides.

3. Draw a circle of radius 3 inches. Place in it a chord AB of length 2 inches. Draw the triangle of greatest area, which is inscribed in the circle on AB as base. Find the area of this triangle.

4. Draw two circles of radii 1 inch and 2 inches whose centres are 3½ inches apart. Draw their four common tangents and measure their lengths.

5. Prove that if two triangles have two sides of the one equal to two sides of the other, each to each, and also the angles contained by those sides equal, the triangles are congruent.

The equal sides AB, AC of an isosceles triangle are produced to D, E respectively, so that AD=AE. Prove that CD=BE.

6. Prove that the square on the hypotenuse of a right-angled triangle is equal to the sum of the squares on the sides.

7. Prove that if a straight line touch a circle and from the point of contact a chord be drawn, the angles which this chord makes with the tangent are equal to the angles in the alternate segments.

Two circles COA, BOD touch at O. Through O two straight lines AOB, COD are drawn to meet the circles, and the chords BD and CA are drawn. Prove that CA is parallel to BD.

8. O is a point within a parallelogram ABCD of which AC, BD are diagonals; show that the sum of the areas of the triangles OAB, OCD is equal to the sum of the areas of the triangles OBC, OAD.

OXFORD LOCAL EXAMINATIONS. JULY, 1911.

PRELIMINARY EXAMINATION.

GEOMETRY. (1¼ hours.)

*[Lines and arcs used in constructions should **not** be rubbed out.]*

1. Define a right-angled triangle.

Draw a triangle PQR such that PQ=3 inches, PR=4 inches and the angle QPR is a right angle. Measure the length of QR.

Bisect QR in S and measure the length of PS. Through Q draw a parallel QT to PR and produce PS to meet it at T. Measure the length of QT.

2. Prove that the angles at the base of an isosceles triangle are equal to one another.

3. Divide a straight line 5 cm. long into three equal parts.

4. Show that the three angles of a triangle are together equal to two right angles.

ABC is an isosceles triangle, having AB=AC and the angle ABC double of the angle BAC ; find the number of degrees in the angles ABC and BAC.

5. Prove that parallelograms on the same base and between the same parallels are equal to one another.

OXFORD LOCAL EXAMINATIONS. JULY, 1911.

PRELIMINARY EXAMINATION.

HIGHER GEOMETRY. (1 hour.)

*[Lines and arcs used in constructions should **not** be rubbed out.]*

1. Show that the sum of the sides of any quadrilateral is greater than the sum of its diagonals.

2. By how much will the top of a ladder 25 feet long, which is standing vertically against the wall of a house, be lowered when the foot of the ladder is drawn 7 feet directly away from the house, while the top rests against the wall?

3. B is a town lying in a straight line between two others A and C, being 5 miles distant from A and 16 miles from C. A fourth town D is 12 miles distant from B and 13 miles from A. Draw an accurate plan, showing the positions of A, B, C, D, and find by measurement the distance from C to D.

4. What is the sum of the angles in all the plane faces of a cube? If the edge of a cube is 5 cm. long, what is the sum of the lengths of all its edges?

5. Show that equal chords in a circle are equidistant from the centre.

OXFORD LOCAL EXAMINATIONS. JULY, 1911.

JUNIOR CANDIDATES.

GEOMETRY. (1½ hours.)

[*No credit will be given for any attempt at a question in Practical Geometry, if any of the construction lines are erased.*]

1. Define a rhombus.

Starting from your definition, prove that the diagonals of a rhombus bisect one another at right angles.

If from the point where the bisector of the vertical angle of a triangle meets the base, straight lines are drawn parallel to the other two sides, prove that the quadrilateral thus formed is a rhombus.

2. Prove that, if in a triangle ABC the side AB is greater than the side AC, then the angle ACB is greater than the angle ABC. (The converse of this theorem must not be assumed.)

Prove that any two sides of a triangle are together greater than twice the median drawn to the middle point of the third side.

3. Divide by a geometrical construction a straight line 7 cm. long into nine equal parts.

4. If the sum of the squares on two sides of a triangle is equal to the square on the third side, prove that the triangle is right-angled.

If the sides of a triangle are 1½ inches, 2 inches and 2½ inches respectively, prove that the square on the perpendicular from the largest angle to the opposite side is 1·44 sq. inches. Construct the square and measure in millimetres the length of a diagonal.

5. Give a construction for drawing a common tangent to two circles which intersect.

If two circles neither touch not intersect, how many common tangents can be drawn to them? Draw illustrative diagrams.

6. Without assuming any property of a circle except the equality of its radii, prove that angles in a segment of a circle which is less than a semi-circle are equal to one another.

OXFORD LOCAL EXAMINATIONS. JULY, 1911.
JUNIOR CANDIDATES.
HIGHER ALGEBRA AND HIGHER GEOMETRY. (1¾ hours.)

Questions 1—4 were on Algebra.

5. If AD is a median of the triangle ABC, prove that
$$AB^2 + AC^2 = 2(AD^2 + BD^2).$$

6. Explain how you could use a protractor and a ruler to draw a tangent to a circle at one extremity of a chord of the circle without drawing any radius of the circle.

If AT is a tangent to a circle and AB is a chord of the circle, prove that the line which bisects the angle BAT will also bisect the arc AB.

7. Prove that the ratio of the areas of similar triangles is equal to the ratio of the squares on corresponding sides.

Show how to draw a straight line parallel to one side of a triangle and bisecting the area of the triangle.

8. AD, BE, CF are the perpendiculars to the sides of the triangle ABC; prove that the circle round DEF will also pass through the middle points of the sides of the triangle ABC.

OXFORD LOCAL EXAMINATIONS. JULY, 1911.
SENIOR CANDIDATES.
GEOMETRY. (1½ hours.)

[*No credit will be given for any attempt at a question in Practical Geometry, if any of the construction lines are erased.*]

1. Prove that the three angles of a triangle are together equal to two right angles.

Find the sum of the interior angles of a polygon of 14 sides.

2. Draw a triangle ABC with sides 3 cm., 4 cm. and 5 cm. Construct a second triangle PQR, so that A, B, C are the middle points of its sides, and prove your construction.

What is the area of the triangle PQR?

3. Enunciate geometrical theorems corresponding to the following algebraical identities:

(1) $(a+b)^2 = a^2 + 2ab + b^2$;

(2) $a^2 - b^2 = (a+b)(a-b)$;

and draw illustrative diagrams.

4. Define a rhombus, and prove from your definition that the diagonals of a rhombus bisect one another at right angles.

Why is it impossible (1) to inscribe a rhombus in a circle, (2) to inscribe a circle in any parallelogram except a square and a rhombus?

5. Construct a regular hexagon about a circle of diameter 6 cm.

Draw also a rectangle equal in area to the hexagon, explaining carefully your method.

6. Prove that angles in the same segment of a circle are equal to one another.

The diagonals of a quadrilateral inscribed in a circle intersect at right angles, and from the point of intersection a perpendicular is drawn on one of the sides. Prove that this perpendicular, if produced backwards, will bisect the opposite side.

OXFORD AND CAMBRIDGE SCHOOLS EXAMINATION BOARD. JULY, 1911.

GEOMETRY. (2½ hours.)

Pass Paper for Higher Certificates.

NOTE. (1) *In answering questions 1, 2, 3 and 10, Candidates must state briefly in words how the problem is solved.*

(2) *In order to pass in Elementary Mathematics, Candidates must satisfy the Examiners in α and β.*

(3) *In order to pass in Additional Mathematics, Candidates must satisfy the Examiners on the whole paper.*

α.

1. In a triangle ABC the side AB is 3 inches long, the side BC is 2 inches long, and the angle BAC, opposite to the side 2 inches long, is 35°. Show by constructing them that there are two distinct triangles which satisfy these conditions; and measure the length of the third side in each of these triangles.

2. Construct a quadrilateral ABCD which satisfies the following conditions. Its area is 4 square inches, the side AB is 2·5 inches long, the side BC is 2 inches long, the angle ABC is 80°, the angle BCD is 126°.

Find, by measurement, its other angles.

3. On a chord 6 cm. long, describe a segment of a circle, the angle in which is 114°.

Construct a parallelogram in which the length of the longer diagonal is 6 cm., the length of the shorter diagonal is 4·5 cm., and the obtuse angle is 114°.

Measure the lengths of its sides.

β.

4. Show that, if two angles of a triangle are equal, the sides opposite to these angles are equal.

5. If a parallelogram is defined in the usual manner as 'a four-sided figure whose opposite sides are parallel,' show that the following figures are parallelograms :

(1) A four-sided figure whose opposite sides are equal.

(2) A four-sided figure in which each pair of opposite angles are equal.

(3) A four-sided figure whose diagonals bisect one another.

6. Prove that parallelograms on the same base and of the same altitude are equal in area.

[N.B. *You are to do this without assuming any formula for the area of a parallelogram or triangle.*]

ABCD is a parallelogram ; P is a point anywhere on BC between B and C, and Q is a point anywhere on CD between C and D. Show that the area of the triangle APQ is less than that of the triangle ABD.

7. Explain and illustrate the geometrical theorem corresponding to the algebraical identity

$$(a+b)^2 = a^2 + 2ab + b^2.$$

ABC is an acute-angled triangle and L is the foot of the perpendicular from A to BC. Show that

$$AB^2 + AC^2 - BC^2 = 2(AL^2 - BL . LC).$$

8. Show that equal chords of a circle are equidistant from the centre; and the converse.

Show that two chords of a given circle can be drawn through a point so as to be of the same given length (less than the diameter of the circle) provided that the point does not lie within a certain concentric circle.

9. Show that, if a straight line touch a circle and from the point of contact a chord be drawn, the angles which this chord makes with the tangent are equal to the angles in the alternate segments.

AB and CD are two straight lines meeting at the point O. Show that the tangents at O to the circles, which pass through OAC and OBD respectively, are inclined to each other at the same angles as the straight lines AC and BD.

γ.

10. Construct, without using the protractor, an equilateral triangle circumscribing a circle of 2 inch radius.

11. Show that if two triangles are equiangular their corresponding sides are proportional; and the converse.

ABCD is a quadrilateral in which the sides AB and CD are parallel but not equal, AC and BD intersect at O. Any straight line through O meets AB at K and CD at L. Show that

KO : OL = AB : CD.

12. Give a method of constructing a mean proportional to two given straight lines, and prove its accuracy.

PQ is a chord of a circle, and the diameter parallel to PQ meets the tangents at P and Q at X and Y. Show that the length of the diameter is a mean proportional to the lengths PQ and XY.

OXFORD AND CAMBRIDGE SCHOOLS EXAMINATION BOARD. JULY, 1911.

GEOMETRY. (2 hours.)

For Lower Certificates.

[*All diagrams should be drawn accurately with a hard pencil. In the questions on Practical Geometry all construction lines should be shown, but no proofs need be given unless they are asked for.*

In order to satisfy the Examiners, Candidates must pass in both the Practical and the Theoretical questions of Part I, and they are recommended not to attempt Part II until they have done all the questions in Part I which they are able to attempt.]

PART I.

Practical Geometry.

1. Draw a straight line AB 2½ inches long and at A and B erect, on the same side of AB, perpendiculars AC, BD 3 inches and 1 inch long respectively. With centre C and radius CD describe a circle cutting AB in E. Measure the length of AE.

2. A cubical block, measuring 3 inches along each edge, is divided into two equal wedges by a plane through two opposite edges. Draw full size one of the faces formed by the division and measure the length of a diagonal of this face.

Theoretical Geometry.

3. If two sides of a triangle be equal, prove that the angles opposite to these sides are equal.

The sides AB, BC of a triangle ABC are bisected at D and E: draw DO, EO perpendicular to AB and BC respectively and let them intersect at O. Show that the angles OAC, OCA are equal.

4. Prove that equal triangles on the same base are of the same altitude, without assuming any expression for the area of a triangle.

Show that a parallelogram can be formed by joining the middle points of the sides of any quadrilateral.

5. Prove that the square on the hypotenuse of a right-angled triangle is equal to the sum of the squares on the other two sides.

PART II.

Practical Geometry.

6. Draw a circle 3 inches in diameter and cut off a segment containing an angle of 50 degrees. Measure the length of the chord on which the segment stands.

7. Construct a square equal in area to a triangle whose sides are 2, 3 and 4 inches.

Theoretical Geometry.

8. If a straight line touch a circle, and if from the point of contact a chord be drawn, prove that the angles which this chord makes with the tangent are equal to the angles in the alternate segments.

If two circles touch and two straight lines are drawn through the point of contact, show that the chords joining their extremities are parallel.

19—2

9. Prove that the internal bisector of an angle of a triangle divides the opposite side internally in the ratio of the sides containing the angle.

In a triangle ABC, D is the middle point of the side BC. DE and DF are drawn bisecting angles ADB, ADC and meeting AB, AC in E and F respectively. Show that EF is parallel to BC.

OXFORD AND CAMBRIDGE SCHOOLS EXAMINATION BOARD. JULY, 1911.

GEOMETRY. (2 *hours.*)

For School Certificates.

NOTE. (1) *In order to pass in Elementary Mathematics, Candidates must satisfy the Examiners in a.*

(2) *In order to pass in Additional Mathematics or to obtain exemption from the Army Qualifying Examination, Candidates must satisfy the Examiners in a and β taken together.*

(3) *Figures should be drawn accurately with a hard pencil, and all 'constructions' clearly shown.*

a.

1. Describe a triangle whose sides are 2·6, 2·8 and 3 inches respectively. Measure the greatest angle.

Inscribe a circle in the triangle; and measure its radius.

2. The area of a parallelogram is 10·5 sq. cm. ; one side is 4·2 cm.; one diagonal is 3·3 cm. From these data two parallelograms of different shapes can be constructed; construct them.

3. Prove that, if two straight lines intersect, the vertically opposite angles are equal.

Show also that the straight lines which bisect adjacent angles are at right angles.

4. Prove that, if two triangles have two sides of the one equal to two sides of the other, each to each, and also the angles contained by these sides equal, the triangles are congruent.

Equilateral triangles CEA, AFB are described outwardly on the sides CA, AB of any triangle ABC; show that BE and CF are equal.

5. Prove that parallelograms on the same base and of the same altitude are equal in area.

[*You are to do this without assuming any formula for the area of a parallelogram or triangle.*]

6. Prove that a straight line drawn from the centre of a circle to bisect a chord, which is not a diameter, is at right angles to the chord; and, conversely, that the perpendicular to a chord from the centre bisects the chord.

ACB is a chord of a circle, C is its middle point, PCQ is another chord through C. Prove that PC is not equal to CQ, unless AB is a diameter.

β.

7. ABCD is a convex quadrilateral; AB is 3 in., BC and CD are each 2 in.; ABC and BCD are right angles. Construct a square equal in area to ABCD.

8. Prove that angles in the same segment of a circle are equal.

A, B are the common points of two circles; C, D are points on the same arc AB of one of these circles. AC, AD meet the other circle again in G, H; BC, BD meet it again in E, F. Prove that the angles EAG, FBH are equal or supplementary.

9. Prove that, if two triangles are equiangular, their corresponding sides are proportional.

D is any point in the side AB of a triangle ABC; DE, parallel to AC, meets BC in E; EF, parallel to CD, meets AB in F. Prove that BF . BA = BD².

UNIVERSITY OF LONDON.

MATRICULATION EXAMINATION.

SEPTEMBER, 1911.

GEOMETRY. (3 *hours.*)

1. O is the middle point of a straight line AB, and C is any point in AB produced. Show that OC is half the sum of AC and BC.

Prove also that the perpendicular from O on any line through C is half the sum of the perpendiculars from A and B on the same line.

Show how the first of these theorems is a particular case of the second.

2. Draw accurately a triangle ABC with its sides AB, BC, CA, 8 cm., 10 cm., and 12 cm. in length respectively. Find a point in its plane at equal distances from B and C, and 7 cm. distant from A.

State the steps of your construction, but do not give a proof of their correctness.

Measure the distances of the point so found from B and C.

3. If two triangles have two sides and the included angle of the one respectively equal to two sides and the included angle of the other, prove that they are equal in all respects.

AB, A′B′ are two equal straight lines. The straight lines bisecting AA′ and BB′ at right angles meet at O. Show that the angle AOB is equal to the angle A′OB′.

4. Explain what is meant by a symmetrical figure.

State (without proof) the axes of symmetry for each of the following figures, illustrating your answers by diagrams (which need not be accurately drawn): (i) isosceles triangle, (ii) rectangle, (iii) rhombus, (iv) semicircle, (v) regular pentagon, (vi) regular hexagon.

5. Give (without proof) any practical test you know of for finding, by drawing and measurement of lengths, whether two given straight lines may be considered parallel.

Show that if all the angles of a hexagon are equal, the opposite sides must be parallel.

6. Show that if the square on one side of a triangle is greater than the sum of the squares on the other two, the angle opposite that side is obtuse.

ABC is any triangle, D a point in the side BC. Show that if AC^2 is greater than $AD^2 + DC^2$, then AB^2 is less than $AD^2 + DB^2$.

Show also why, if D is the middle point of BC, the excess in the one case is equal to the deficiency in the other.

7. Show that equal chords of a circle are equally distant from the centre.

If two equal chords of a circle, AB, CD, intersect at O, show that AO is equal either to OD or else to OC.

8. Show that if a quadrilateral inscribed in a circle has one of its sides produced, the exterior angle so formed is equal to the interior and opposite angle of the quadrilateral.

K is the middle point of the arc BC of the circle circumscribed to a given triangle ABC. KL, KM are drawn perpendicular to AB, AC. Show that LB = MC.

9. Draw carefully an equilateral octagon, with each of its eight sides distant 1·5 inch from a given point. State the steps of your construction, but do not give a proof of their correctness. Measure the length of each side of the octagon.

10. The inscribed circle of a square ABCD touches the side AB at E. From the middle point of AE a tangent is drawn to the circle and is produced to meet CD produced in F. Show that DF is equal to the radius of the circle.

COMMON EXAMINATION FOR ENTRANCE TO PUBLIC SCHOOLS.

NOVEMBER, 1911.

GEOMETRY. (50 *minutes*.)

Do any three parts marked (a) and any three parts marked (b). The later questions are marked more highly than the earlier ones.

[*Accurate figures are not required except in the question marked thus*.]

1. *(a) Draw a triangle ABC having AB=3·2 in., ∠BAC=50°, ∠ABC=65°. How big is the angle ACB? What could you foretell about the sides of the triangle?

(b) Draw a parallelogram; call it ABCD, A being at an acute angle; join BD. Assuming that ∠BAD=50° and ∠ABD=60°, calculate each angle in the figure. Give your reasons carefully.

2. (a) ABC is an isosceles triangle having AB=AC; AD is drawn perpendicular to BC and meets it at D; prove that D is the mid-point of BC.

(b) PQR is an isosceles triangle having PQ=PR. A straight line is drawn perpendicular to QR and cuts PQ, PR (one of them produced) in X, Y. Prove that the triangle PXY is isosceles.

3. (a) Show how to draw the perpendicular bisector of a given straight line, using only a straight edge and compasses. State your construction and give a proof.

(b) ABCD and ABPQ are a parallelogram and a rectangle on opposite sides of a straight line AB; join DQ, CP: prove that CDQP is a parallelogram.

4. (a) Prove that triangles on the same base and between the same parallels are equal in area.

(b) FGH is a triangle, K is the mid-point of GH, and P is any point on FK; prove that the triangles FPG, FPH are equal in area.

5. (a) Explain how to draw a circle through three given points. Give a proof.

(b) A triangle ABC is right-angled at A, and D is any point in AB; prove that $CB^2 - CD^2 = AB^2 - AD^2$.

6. (*a*) Prove that angles in the same segment of a circle are equal to one another.

(*b*) Draw a quadrilateral ABCD and its diagonals. Now suppose that a circle could be drawn through the points A, B, C, D and that ∠DAC=48°, ∠BDC=52° and ∠BDA=64°, calculate the other angles in the figure. Give your reasons carefully.

FOR NAVAL CADETSHIPS. JULY, 1911.

GEOMETRY. (1½ *hours.*)

[*In questions 1—4 all lines used in the constructions must be shown, but no descriptions of the constructions need be written except in question* 4.]

1. Draw a triangle ABC having BC=6 cm., ABC=35°, ACB=110°. Construct a perpendicular from A to meet BC produced in D. Measure AD.

2. Construct an angle of 22½°, using only the ruler and compasses.

3. Construct a square of side 1·9 inches, draw the diagonals and measure them.

4. Through a given point construct a parallel to a given straight line, describing your construction in words.

---

5. Define *angle, right angle, parallelogram.*

6. If two angles of a triangle are equal what is known about the sides?

In a triangle ABC the angles ABC, ACB are equal and the lines BO, CO bisect them, meeting in O; prove that OB=OC.

7. ABC is an angle of 60° and on the side of AB remote from C angles ABD=119° and ABE=130° are drawn. Of the two bent lines CBD, CBE which is more nearly straight, and why?

8. What is known about the sum of the angles of a triangle?

If one angle of a triangle is greater than the sum of the other two, prove that the triangle is obtuse-angled.

INDEX—LIST OF DEFINITIONS.

Acute angle, obtuse angle, reflex angle. An angle less than a right angle is said to be acute; an angle greater than a right angle and less than two right angles is said to be obtuse (p. 16); an angle greater than two and less than four right angles is said to be reflex. (p. 179.)

Acute-angled triangle. A triangle which has all its angles acute is called an acute-angled triangle. (p. 35.)

Adjacent angles. When three straight lines are drawn from a point, if one of them is regarded as lying between the other two, the angles which this line makes with the other two are called adjacent angles. (p. 25.)

Alternate angles. (p. 30.)

Altitude. See **triangle, parallelogram**.

Ambiguous case. (p. 85.)

Angle. When two straight lines are drawn from a point, they are said to form, or contain, an angle. The point is called the **vertex** of the angle, and the straight lines are called the **arms** of the angle. (p. 13.)

Angle in a segment. An angle in a segment of a circle is the angle subtended by the chord of the segment at a point on the arc. (p. 182.)

Angle of elevation, of depression. (p. 68.)

Apollonius' circle. (p. 250.)

Apollonius' theorem. (p. 141.)

Arc of a circle. (p. 150.)

Area of circle. (p. 202.)

Base. See **triangle, parallelogram**.

Bisect. (p. 25.)

Change of a figure. (p. 86.)

Chord of a circle. (pp. 53, 150.)

Circle. A circle is a line, lying in a plane, such that all points in the line are equidistant from a certain fixed point, called the **centre** of the circle. The fixed distance is called the **radius** of the circle. (p. 149.)

Circumcentre. The centre of a circle circumscribed about a triangle is called the circumcentre of the triangle. (p. 155.)

Circumference of a circle. (pp. 149, 160.)

Solid. (p. 1.)

Square. A rectangle which has two adjacent sides equal is called a square. (pp. 38, 90.)

Straight line. (p. 2.)

Supplementary angles. When the sum of two angles is equal to two right angles, each is called the supplement of the other, or is said to be supplementary to the other. (p. 24.)

Surface. (p. 1.)

Symmetry. (pp. 106, 151.)

Table of facts or theorems. (p. 75.)

Tangent. A tangent to a circle is a straight line which, however far it may be produced, has one point, and one only, in common with the circle. (pp. 166, 167.)

The tangent is said to **touch** the circle; the common point is called the **point of contact.**

Tetrahedron. A solid with four triangular faces.

Third proportional. If x is such a magnitude that $a : b = b : x$, then x is called the third proportional to the two magnitudes a, b. (p. 224.)

Trapezium. A quadrilateral which has only one pair of sides parallel is called a trapezium. A trapezium in which the sides that are not parallel are equal is called an **isosceles** trapezium. (p. 90.)

Triangle. A plane figure bounded by three straight lines is called a triangle. (p. 33.)

Any side of a triangle may be taken as **base.** The line drawn perpendicular to the base from the opposite vertex is called the **height,** or **altitude.** (p. 120.)

The straight line joining a vertex of a triangle to the mid-point of the opposite side is called a **median.** (p. 61.)

Trisection of an angle. (p. 109.)

Vertically opposite angles. The opposite angles made by two intersecting straight lines are called vertically opposite angles (*vertically* opposite because they have the same vertex). (p. 27.)

Vertices. The corners of a triangle or polygon are called its vertices. (p. 37.)

CAMBRIDGE: PRINTED BY JOHN CLAY, M.A. AT THE UNIVERSITY PRESS

GODFREY AND SIDDONS'S GEOMETRY.

This work is published in the following forms:—

(1) Complete in One Volume, Large Crown 8vo. 3s. 6d.
Or together with Solid Geometry. 4s. 6d.
Or (2) in Two Volumes.
 Vol. I. (Experimental Course, and Books I. and II.) 2s.
 Vol. II. (Books III. and IV.) 2s.
Or (3) in Five Parts:—
 Part I. Experimental Geometry. 1s.
 Part II. Theoretical Geometry. Book I. 1s.
 Part II.   ,,   ,,   Book II.—Area. 1s.
 Part II.   ,,   ,,   Book III.—The Circle. 1s.
 Part II.   ,,   ,,   Book IV.—Similarity. 1s.
 Part II. Theoretical Geometry is also published as a separate volume. 3s.

ANSWERS TO THE EXAMPLES. 4d. post-free.
SOLUTIONS OF THE EXERCISES. By E. A. PRICE, B.A., Master
 at Winchester College. Crown 8vo. 5s. net.

By the same Authors.

Geometry for Beginners. Crown 8vo. 1s.

This book adopts the suggestions on the teaching of Geometry to beginners
contained in the Board of Education circular of 1909.

NOTES AND ANSWERS TO EXERCISES. 4d. post-free.

Modern Geometry. Crown 8vo. 4s. 6d.

A sequel to 'Elementary Geometry, Practical and Theoretical.'

Solid Geometry. Crown 8vo. 1s. 6d.

www.ingramcontent.com/pod-product-compliance
Lightning Source LLC
Chambersburg PA
CBHW081106170526
45165CB00008B/2344